HOW TO F**KING SAVE THE PLANET

Published in 2021 by Welbeck

An imprint of Welbeck Non-Fiction Limited,
part of Welbeck Publishing Group.

20 Mortimer Street London W1T 3JW

Logo © IFLScience!. All rights reserved.

Text © Welbeck Non-Fiction Limited, part of Welbeck Publishing Group.

A CIP catalogue record for this book is available from the British Library

ISBN 978 1 78739 432 2

Printed in Dubai

10 9 8 7 6 5 4 3 2 1

JENNIFER CROUCH

HOW TO F**#KING SAVE THE PLANET

THE BRIGHTER SIDE OF THE FIGHT AGAINST CLIMATE CHANGE

WELBECK

CONTENTS

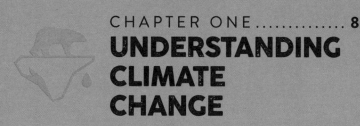

INTRODUCTION

THIS BOOK CANNOT SAVE THE PLANET. No single person or object or plan or approach can do that, because (as many of us know) a vast amount of damage has already been done to the planet. The nature of this anthropogenic (human-caused) damage is complex, and it does not affect all organisms equally. It has also been accumulating since the 1450s – so we're going to need something more than some gaffer tape and a can-do attitude.

What does it even mean to save (or lose) a planet? What is being lost, and how? What have we done? What are we actually doing? Who are the 'we' who have made this mess? Are 'we' all equally responsible? Who can (and should) help, and how?

Planet Earth is experiencing a process of extreme loss that will become even more extreme if we do not change our behaviour as well as the systems that serve and encourage environmental destruction. Since the industrial revolution, climate damage has been accelerating dramatically – particularly from the 1950s onwards. Due to pollution, greenhouse gases, habitat destruction, increasing global average temperatures and general greed we are already losing coastal cities, ruining the health of the oceans and destroying our jungles, forests and ice sheets. Deserts are growing, cities keep expanding and plants and animals cannot keep up with the changes to which they are being subjected. Anthropogenic industrial transformation of the land, air and water is causing mass extinction on a vast scale – in ways that have never happened before. Human lives are also being lost due to drought, wildfires, flooding, inequality, conflict over resources, and increasing insecurity. Not to mention the trauma and anxiety people feel because of all this. What is the endgame of this process of loss? How much can the planet take? How much can we tolerate?

Who has the power to do anything about the total sh*t show we see unfolding before us? Climate change is a complex crisis that expresses itself in multiple interrelated ways. This book presents a mere sliver of the knowledge and research about climate change and its extent. At the back you can find a list of resources and further reading to help you explore things further and describe some of the things that we are doing wrong and how they might begin to be changed.

Our resounding cry is that now is the time to continue efforts for change. So many have been calling for action for a long time – tirelessly campaigning, protesting and resisting. Climate destruction has got to a point where we will (regardless of what happens next) have no choice but to change, *drastically*. Things simply cannot continue as they are. Things cannot be allowed to progress into more misery, extraction, domination and exploitation. The strategy of infinite growth at the expense of all that surrounds is literally the strategy of cancer. The key question is – what choices can we make now to improve the land, air, water and lives of all those who exist on planet Earth?

Currently, we are not all equally empowered to take action – but we can all listen to each other, learn to understand and support each other, protect each other and hold those responsible to account by challenging and changing the current ideologies, processes and methods that serve the top few percent.

How hard could it be?!

CHAPTER ONE

UNDERSTANDING CLIMATE CHANGE

WHAT IS CLIMATE CHANGE?

OUR PLANET IS SCREWED.

We are literally flying through space on a giant-ass rock and we are attacking, polluting and all-round-screwing-over the very thing that keeps us alive in the universe.

Every single living thing is connected. We don't mean that in a woo-woo New Age way; it's scientific fact. The Earth's oceans, currents, land masses, water, animals and people, inside and out, are all intricately interconnected: we all occupy the biosphere together. Just as one example, seasonal weather cycles that we have (or should have!) today and animal migration patterns have co-evolved over millions and millions of years – gradually and stably enough for complex, diverse ecosystems to evolve and adapt within them.

Then humans came along and messed it up. Climate modelling suggests that there is a 97% to 100% chance that the increase in the global average temperature is caused by human industrial activity. Scientists usually save this level of certainty for statements such as: "grass is real", "there is a sun" and "kebabs taste better at exactly 3.16am". We have caused large-scale, dramatic and consistent accumulative changes to the environment that have disrupted these interconnected cycles like nothing that has ever been seen before. We're now closer to the brink than we might like to believe.

There is also a 95% chance that the Earth's average temperature will increase by more than 2 °C this century. "Two degrees?" you might say? That's not too bad! It can fluctuate two degrees a day, even an hour, without anyone batting an eyelid. No, no... this is very, DISASTROUSLY, EXTRAORDINARILY bad. There is a huge difference between weather and average global temperature. Weather changes daily but average global temperature remains stable for millennia.

Global average temperature is the measure of the average global surface temperature. Detailed, accurate historical measurements date from 1880 (with some going as far back as 1850), when the first thermometer-based records began. Based on readings collected way back when, global average temperature was given a baseline of 14°C. 2020 was the hottest year on record. Ice cores show that it normally takes tens of thousands of years for this average temperature to fluctuate even by a single degree. Two degrees difference in global average temperature over a couple of hundred years is enough to completely destroy the Earth's natural cycles to such an extent that it will, without a doubt, cause serious disruption and destruction to life as we know it.

SLOW YOUR ROLL! WHAT DAMAGE CAN TWO DEGREES REALLY DO?

- Ice caps will melt causing sea level to rise; this will cause island nations and coastal regions to become more flood-prone and some will even be submerged.
- It will change the distribution of rain (or precipitation if you want to get technical) to become more extreme; wet areas will become wetter, and dry areas drier.
- Heatwaves and droughts will increase, threatening food security and health, and increasing desertification.
- As global temperatures increase, the surface temperature of the ocean does too, which increases the intensity of storms and hurricanes.
- Some hot and humid areas could become too hot to live in.

And that's only the tip of the melting iceberg. The quality of the air we breathe, the water we drink and the food we eat is already being affected by these changes all over the globe. Much like Pitbull, climate change really is Mr Worldwide.

Of course there are natural changes that occur in the climate such as; the effects of the Sun, movement of water vapour, wildfires, changes resulting from the seasons, and geological activity such as volcanoes. But humans have taken these factors and added our own heavily fertilised crap to the mix: greenhouse gases from industry, pollution and waste from our homes, industries, transport systems and cities, discarded products such as plastics, electronics and chemicals which build up in the air, soil and water. Natural change exists, but not nearly to the scale that we're seeing today.

"But if it's called global *warming*, why is it getting *colder* as well?"

Excellent question, random audience member. Increasing global temperatures creates more energy in Earth's systems. So, a better way to describe global warming is that the Earth's systems are actually becoming more energetic (rather than simply warmer). This increase in energy then causes weather and natural climate changes to become far more dramatic and extreme, resulting in more extreme variations in weather. The climate can be thought of as a toddler that you've fed too much sugar to and is now running head-first into a flatscreen TV.

Everything, and I mean everything, comes back to climate. Climate is **long-term weather patterns and cycles**. This includes things like the interactions between heat transfer in water, intensity of storms, erosion, activity in the atmosphere and ocean currents. These are all affected by and part of the climate.

"IN CASE WE HAVE FORGOTTEN, BECAUSE WE KEEP HEARING THAT 2014 HAS BEEN THE WARMEST YEAR ON RECORD, I ASKED THE CHAIR, DO YOU KNOW WHAT THIS IS? IT'S A SNOWBALL JUST FROM OUTSIDE HERE. SO IT'S VERY, VERY COLD OUT. VERY UNSEASONABLE."

In 2014, US Senator Jim Inhofe, the Republican chairman of the Environment and Public Works Committee made headlines by chucking a snowball in a debate on climate change, clearly not understanding the difference between climate change and errrr, weather.

THE ANTHROPOCENE

We're living in the Anthropocene. Surprise!

Okay, so, what is the Anthropocene? According to some geologists, the Anthropocene Epoch is an unofficial division of geologic time used to describe the most recent period in Earth's history, when human activity started to have a significant impact on the planet's climate and ecosystems.

The term "Anthropocene" is becoming more and more popular in climate conversations. First popularised by atmospheric chemist Paul Crutzen and limnologist Eugene Stoermer in 2000, when they combined the Greek words for human (*anthropos*) and recent (*kaino*) to describe the current period in geological history when human activity is so extensive that it is causing actual geological changes to occur around the world. Many of these changes will persist and continue to affect the Earth for millennia. Cruzten and Stoermer argue that the term is necessary because human industry and pollution are already having permanent effects on geology and forming a distinctive "human-made" or "industry-made" geological stratum that is now accumulating in the Earth's surface

Not all geologists are convinced by this yet, and so the Anthropocene is not currently a formal geological division in the Geological Time Scale. Officially, our epoch is categorised as the Meghalayan Age of the Holocene Epoch (whip that out in a pub quiz and you'll look hella smart), but some geologists are currently proposing to scientifically formalise the Anthropocene.

Some critics argue that the term Anthropocene is not the best because the term *anthropos* identifies *all humans* as part of the problem. But, realistically, that's not the case. There are plenty of Indigenous and rural communities with methods of fishing, farming, building and making that don't devastate the environment. YOU also didn't ASK for all those layers of plastic packaging on that thing you ordered online so who decided *for you* to wrap it up in a million layers of non-recyclable bubble wrap?

Hate to break it to you mate, but it's the current lifestyles of wealthy countries that have caused the Anthropocene. Consumerism and the scale of consumption that's part of contemporary capitalism is the main cause of the excessive damage we see. For this reason, the term "Capitalocene" is sometimes used instead to link anthropogenic climate change to specific behaviours like exploitation, profit at the cost of all life, and extractive practices. Instead of looking at climate change as a by-product of humanity, we're far better off addressing it as a result of our collective behaviour. Other proposed terms for the Anthropocece include Plantationocene, Chthulucene and Misanthropocene.

Ultimately, these names are all trying to show the same things: we need to make the future more liveable by fundamentally changing our behaviour, especially in the rich western world. It's called accountability, Google it.

....

"THE HOLOCENE HAS ENDED. THE GARDEN OF EDEN IS NO MORE. WE HAVE CHANGED THE WORLD SO MUCH THAT SCIENTISTS SAY WE ARE IN A NEW GEOLOGICAL AGE: THE ANTHROPOCENE, THE AGE OF HUMANS."

―――――――――

Sir David Attenborough, the conservationist, the legend, the man.

....

CLIMATE SCIENCE AND OTHER SMART STUFF

Reason number 675 why we should pay attention to climate change: the people who are telling us to worry about it are super smart and have dedicated their lives to understanding climate. No one WANTS this to be a disaster, that's just what experts keep finding. Anyone who has discussed climate change online will know that there are a lot of people who either claim to be climate scientists, claim to know as much as climate scientists, or claim climate scientists don't know a thing (let's not get started on the ones who think climate scientists are all part of a secret organisation of lizard people trying to trick us sheeple into believing in climate change for... reasons??).

So, it's important, therefore, to know what makes a climate scientist, why they are qualified to say what they are saying and, if you're interested, even how to become one!

Scientists don't just sit around all day in lab coats. That's only part of the day, usually around elevenses. When scientists study our world, they measure and record observations of it using numerous monitoring devices such as satellites, weather radars, and by using samples to track changes in soil, water and atmospheric purity. Scientists track a whole range of changes from temperature, humidity, to the quality of water and the balance of different gases in the atmosphere. They then bring these different observations together and combine them with existing knowledge to create a theory of what is happening. So, yeah, tell me again how a Twitter troll who lives in his mother's basement knows more about climate change than actual scientists?

Very simply, a climate scientist is a person who studies and has expert knowledge of one or more of the natural or physical sciences. Climatologists use analysis of observations and modelling of physical laws that determine the climate in order to understand the world and predict what might happen in the future. There are many types of climate scientists with all sorts of specialities. These include, but aren't limited to; biodiversity, climate systems, geophysics and climatology.

By paying close attention to environmental indicators, climate scientists can track the intensity and activity of weather systems, ocean currents and temperatures. This is data that actively impacts the lives of all species on the planet, including humans. How these environmental indicators are increasing or decreasing is vital in our understanding of how our climate is changing and becoming more energetic. So, what's changing? Surface air temperature, humidity, ocean temperature and land temperature are all increasing, while sea ice, glaciers and snow cover are all decreasing. None of this is good... and that's putting it very lightly...

"...97 PERCENT OR MORE OF ACTIVELY PUBLISHING CLIMATE SCIENTISTS AGREE: CLIMATE-WARMING TRENDS OVER THE PAST CENTURY ARE VERY LIKELY DUE TO HUMAN ACTIVITIES."

NASA, 2020. Yes, the actual NASA. The people who just landed ANOTHER rover on Mars.

GLOBAL CIRCULATION
MODELS (GCMS)

If climate scientists were superheroes (and we think they are) then global circulation models (GCMs) would be their trusty sidekicks. Climate scientists use evidence of Earth's past climate that is embedded in our surroundings. Tree rings, ice cores, fossils and geology all hold physical and chemical signatures of the past. Analysing these helps scientists to figure out what the climate was probably like millions of years ago, giving climate scientists the data to make the predictions that they do. Their research, kinda unsurprisingly, is a lot more rigorous and detailed than your uncle – the keyboard warrior – who watched a 15-minute YouTube video and now apparently knows more than the UN.

Using this data, scientists are able to create super-complex mathematical reconstructions of our environment, which are called GCMs. These clever little models are designed to mimic the atmosphere, biogeochemical cycles, ocean currents and global average temperature change. These computer models are **mind-bogglingly complicated**, but it's fair to ask how we can trust something that relies so much on predictions.

However, the answer is as simple as GCMs are complex: they are consistently held up to great scrutiny. They are first tested by running current measurements through them and comparing the results that they spew out with observations from the real world. Then a second test of the accuracy of a model is "hindcasting" – using them to predict what the climate was like in the past. This can reliably be carried out as far back in history as there are meteorological records (the 1850s or so). Showing that our models **can accurately predict the past *and* the present** can give us far more confidence that they will do the same for the future. How cool is that?

"[MODELS] ARE FULL OF FUDGE FACTORS THAT ARE FITTED TO THE EXISTING CLIMATE, SO THE MODELS MORE OR LESS AGREE WITH THE OBSERVED DATA. BUT THERE IS NO REASON TO BELIEVE THAT THE SAME FUDGE FACTORS WOULD GIVE THE RIGHT BEHAVIOUR IN A WORLD WITH DIFFERENT CHEMISTRY, FOR EXAMPLE IN A WORLD WITH INCREASED CO2 IN THE ATMOSPHERE."

Freeman Dyson, an great physicist, but terrible climatologist. For some reason, he believed that we could genetically engineer trees to grow on comets, but couldn't quite believe we could model climate successfully...

HOW TO MAKE YOUR OWN GCM

1. The first extremely simple step: create software that can model the world. It must contain an accurate (numerical) 3D map of the world divided into squares – called "grid cells" – measuring just a few hundred kilometres each. These must cover land, sea and air. Easy, right?
2. Next, separate data must be inputted into each cell, which covers all of the climatic measurements that we have for that area. The smaller the cell and the more data you have, the more accurate and granular your GCM can be. Pfft, anyone can do that.
3. Now you must write the code to model the Earth's conditions. This will include equations of motion, thermodynamics and fluid dynamics, as well as metrics such as temperature and wind speed. Typically this code runs to tens of thousands of pages. All in an afternoon's work?
4. Finally, click run and wait! However, unless you have a supercomputer roughly the size of a tennis court, you might be waiting a while... Maybe this is more difficult than we thought...

HOW ACCURATE ARE GCMS?

More accurate than your horoscope, that's for sure. No, Mercury wasn't in retrograde, you were just having a bad day. It happens. GCMs, though, are becoming increasingly accurate. Just as one small example, in 2012, scientists started to include equations that simulated how water behaved in the Earth's cooler polar regions. This information was fed back into new GCMs, which in 2013 provided data that more accurately matched observed data. This is exactly how these things are meant to work.

However, it is important to note that climatic systems are far too complex to be perfectly simulated by any computer. We **should not expect any**

GCM to be correct 100 per cent of the time but, similarly, just because a GCM is incorrect about one thing does not mean that all predictions are! This is something deniers would **love** to do, even when Saturn is rising.

As more and more independent studies using different GCMs and studies of real-world situations are carried out, and the analyses continually yield the same conclusions – which they do – then we can say that our predictions are reliable. Predictions of increasing global temperatures have been a consistent feature of all significant GCMs since 1975, and have been borne out by actual temperature measurements. BOOM!

The majority of the world's leading science organisations have issued statements explicitly affirming that climate change is happening. Also, a 2013 publication by the Intergovernmental Panel on Climate Change (IPCC) claimed a 97.1% scientific consensus based on the abstracts of 11,944 scientific papers on climate published between 1991 and 2011. Climate change is real, people! Even if there are a few crazy (or fossil-fuel-funded!) scientists out there willing to say it's not.

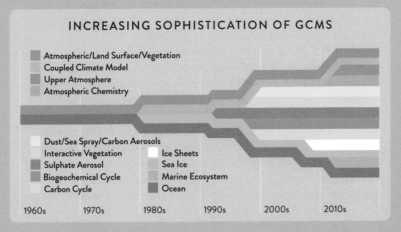

INCREASING SOPHISTICATION OF GCMS

- Atmospheric/Land Surface/Vegetation
- Coupled Climate Model
- Upper Atmosphere
- Atmospheric Chemistry

- Dust/Sea Spray/Carbon Aerosols
- Interactive Vegetation
- Sulphate Aerosol
- Biogeochemical Cycle
- Carbon Cycle
- Ice Sheets
- Sea Ice
- Marine Ecosystem
- Ocean

1960s   1970s   1980s   1990s   2000s   2010s

DENIER 101:
THE PSYCHOLOGY OF
CLIMATE CHANGE

Look, things suck right now. All you have to do is open Twitter and you will be bombarded with everything from bad news to terrible politicians. In amongst the cat memes and witty one-liners, you can see constant information on mass extinction, the death of the coral reefs, COVID-19, increasing flooding, the rise of authoritarianism and fascism... the internet is a cheery place to be. It's no wonder that it's hard to find anything to be optimistic about.

But, this is it now, it's OUR generation that has to demand a better future. If we don't fix the crap that we've been saddled with by previous generations – knowingly and unknowingly – then we're no better than them. In fact, we would be worse, because there's no way that we can say we didn't know the consequences. It is ultimately our choices, our votes and our behaviour that can prevent the scale from tipping too far, and, you know... a complete bloody climate meltdown.

We get it, it's hard to act on issues that seem far away, either time-wise or geographically. It is difficult to imagine people in the USA agreeing to shut down the economy in order to save the lives of people in Bangladesh. Or imagine people in the UK or Europe acting in sufficient numbers if they were told that if they all washed their hands and wore a mask in public spaces they would save the lives of people in Mauritius or Indonesia. It's not easy to act on even simple interventions that would save the lives of people on the other side of the world when you can't see the effects with your own eyes. It's even hard to care about the polar bears when you have people to see and things to do (or things to see and people to do, wahey!). Part of the human condition includes finding it hard to admit we are wrong, and hard to change. It's only human to struggle to take that empathetic leap and turn doomscrolling into action..

However, for this to work, our care has to extend beyond ourselves and our loved ones. It has to extend all the way to people in other parts of the world and even to people who don't exist yet. Those who live in other countries or who are yet to be born are **just as important** as those who live in your neighbourhood or are part of your family. We're all part of that big Earth village, dude. We need to be moved EVERY GODDAMN DAY to fix this massive stinking mess.

Now we've got that settled, depending on what end of the spectrum you're on, your feelings about climate change probably fall somewhere in between eco-anxiety and despairing denial.

ECO-ANXIETY

With over a million species on plants and animals on the brink of extinction, and a predicted billion climate migrants (one BILLION!) by 2050, ecosystem collapse imminent and the destruction of vital eco-infrastructure... who wouldn't be having nightmares? Research shows that more than 70% of 18- to 24-year-olds in the UK have experienced "eco-anxiety" (loosely defined as anxiety about the future and climate collapse). This is a difficult problem to tackle, and while education on climate change is getting better, we must also make sure that education includes the solutions. It's also important to think about climate change holistically. This can be done by bringing education outside of the classroom to look at the wider issues – particularly climate racism and the issues surrounding free-market capitalism and climate change.

CLIMATE PSYCHOLOGY

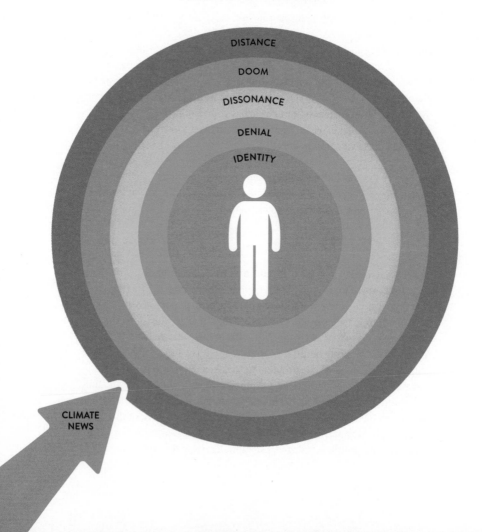

When we see images of the Amazon burning, hear of mass extinctions or wars over resources, read news coverage about wildfires, floods and disasters and their human and non-human casualties, it is hard not to feel helpless. Helplessness is not a flaw! If you feel sad it's because you understand what is actually happening, gold star to you. Have a good cry or a scream into your pillow or throw a plate at a wall, go on, we'll wait...

Feeling better? Good, now take all that hopelessness and despair and sadness and crush it into a hard red ball of anger. Now is the time for strength, to bloody do something about it. Can you do better in your daily choices? Do you *really* need that crap off Amazon? Is it really going to take that much time out of your day to just fricking recycle? Are you sure you know about the people you're voting for? This is *not* all your problem to fix – as much as Big Oil wants you to think it is. But by starting to live and vote for the changes we want to see, small steps can lead to sweeping systemic change.

DENIAL

Here's the thing with denial, you don't have to be a complete hard-headed climate change denier to have doubts. But, unfortunately, we are now at the stage where to not actively fight against climate change is a form of denial. When was the last time you had an intruder stalking through your house and you let them roam around for a bit in the hopes that your kids would resolve the situation, one way or the other?

Even if you strongly believe that climate change is an important issue, you could still be partaking in behaviour known as "soft denial". This is when you know something is wrong, but emotionally it doesn't affect you and you don't do much to stop it, like frequently using air travel, wasting food, eating steak every night, or putting the heating on in the springtime (your Dad was right, put on another jumper, what do you think I am, made of money?!).

When you don't see the day-to-day effects of climate change it can be easy to ignore. If it doesn't fit your world view, it can be even easier to dismiss, especially if you don't take the time to educate yourself. A 2019 study from the Yale Program on Climate Change Communication found that about 16% of Americans don't think global warming is happening at all. Almost seven in ten Americans do believe that climate change is happening. Of this 68% that do have some common sense, only 46% are "extremely" or "very" sure it is happening, most believe that it is at least partly human-caused, but few are engaged with climate change emotionally, behaviourally, or politically, even though 38% claim to have been personally affected by it, a key example of soft denial.

In other parts of the world, it's a whole different story with 88% to 97% of those interviewed as part of the European Social Survey published in 2018 believing that climate change is happening and agreeing that climate change is caused by human activity – oh, and that the impacts of it will be bad for everyone and everything (how very dramatic). However, in the UK, Ireland, the Netherlands, Estonia, Belgium and Finland (shockingly) only 59%-66% believe the impact of climate change will be bad.

A 2019 study by Afrobarometer found that of people interviewed on the African continent, 85%-79% report worsening weather conditions for farmers, 63% of people considered climate change to be a very serious problem with 52% believing that climate change is caused by human activity. 67% reporting that climate change is making lives in their countries worse, 71% saying that it must be stopped, and 51% of respondents arguing that ordinary people can do a least "their bit" to help.

A 2017 national Chinese survey published by China's Centre for Climate Change Communication found that 94.4% said that climate change is happening, 66% believed it was being caused by human activities, and 80% are worried about it. So here's the thing, the majority of people believe that climate change is happening and is bloody awful, but how many of those surveyed are actually *doing* something about it? Not that many, unfortunately.

That's where this book – and others like it – come in, of course (nudge, nudge). We've established that soft denial isn't great, but at least people are *aware* of the issues and agree that change is needed. For these people, the key is to show how what we can do in our day-to-day lives can make a difference.

But the next level is where your full-blown, card-carrying, hard-as-hell deniers enter the scene. They are the real deal. At its heart, climate denier-ism is an attempt to protect a way of life. Deniers may be genuinely convinced that everything is fine, that their way of life cannot and should not change, that what humans do can't actually change anything, or that, ultimately, it doesn't matter what happens – at least, it doesn't matter to them. Usually, but not always, they tend to be people who are not directly affected by climate change or are privileged enough to ignore the signs of climate change in their lives, which is why it doesn't matter to them. It's very easy to ignore wildfires when they aren't burning *your* home. In some extreme cases, an acknowledgement of climate change is a threat to their very identity (see more on this on page 30).

HOW TO TALK ABOUT CLIMATE CHANGE ON SOCIAL MEDIA

So you want to spread the word about climate change in your friend group or on social media? Maybe even change a few hearts and minds in the process? Follow these handy tips to keep yourself safe, sane and sassy.

BUILD YOUR COMMUNITY

Who are you actually talking to, and do they give a damn about what you've got to say? Can this community reach out to individuals you know that are on the fence or in denial about climate change? Fostering communities of mutual understanding and love that listen to each other, empathise with and believe each other is important for self-preservation, mutual support and encouragement, but your climate journey does not stop there.

WHO ARE YOU CENTERING?

Climate racism – and the fact that white middle-class liberals have not centered Black, Brown and Indigenous voices in the climate change narrative – is part of the problem. Are you engaging with communities outside of your immediate surroundings? Are you centering the voices of Indigenous, Black and Brown communities? Are you platforming the poorest, worst hit and most vulnerable communities? Climate change cannot be centered around expensive sustainable products or white people doing yoga. When marginalised people speak, listen to them about their experience. It's not up to marginalised individuals to do all the work of winning deniers over.

DON'T GO FOR THE WIN

Are you desperate to tell people what you think? You already know what you think! So it's important to listen to each other if we have any hope of really trying to win people over. Look, we all love a good viral moment but real conversations are about genuine exchange and entering a conversation. Thinking about how you are "going to win" is more likely to escalate into conflict which will not resolve these complex and messy issues or win over anyone on the other side of the fence.

"I HAVE A DREAM THAT THE PEOPLE IN POWER, AS WELL AS THE MEDIA, START TREATING THIS CRISIS LIKE THE EXISTENTIAL EMERGENCY IT IS."

Greta Thunberg, all-round awesome
environmental activist, 2019.

HOW TO TALK TO CLIMATE CHANGE DENIERS

Climate change deniers tend to fall into a few broad categories. Figuring out which type of denier you're dealing with can be super-helpful in starting a conversation, and hopefully changing their minds. Here's the breakdown of the main denier flavours:

The Flat-Out Denier: "Climate change is not happening."
The Partial Denier: "Pollution is happening but warming isn't."
The Watched-*Ice-Age*-Once Denier: "What we are seeing is natural climate change."
The Misplaced Optimist: "It is happening, but it's not really that bad."
The Greedy Banker: "The economy is more important."
The Keyboard Scientist: "Humans are incapable of altering the climate, the whole concept is ridiculous!"
The Nihilist: "We're all gonna die anyway, may as well go out reeking of Lynx."

Buckle in guys, cause we're gonna teach you how to talk to deniers! Okay, so, picture this: your uncle has been sharing some really dodgy stuff on Facebook and you're starting to worry that he might be a climate denier. You're sitting next to him at Grandma's 80th birthday party. Do you:

a) Ignore his climate denial ramblings and Tweet about it secretly.
b) Get into a shouty argument about emission reduction curves and upset Grandma.
c) Find common ground to engage him on and slowly convince him to consider changing his stance.

Correct, the answer is c, gold star for you.

Firstly, get personal. Talking about your own experiences of taking part in community clean ups, or about your love for your local area and how it has changed over the last 5, 10 or 20 years can really break the ice. Talk about something tangible and real world that he can picture.

Secondly, find common ground. Everyone will have an experience of climate change – even if they are a denier. Maybe your uncle enjoys fishing, and perhaps he's noticed changes in fish stocks or water quality? Maybe he's noticed that there is more flooding than there used to be? This could provide a safe starting point for a positive conversation.

Thirdly, meet them where they're at. Your uncle is probably a conservative voter. Such a demographic usually responds better to notions of landscape, preservation, continuity, tradition and protection. Making the link between these values and conservation is a great starting point into moving him away from more radical viewpoints.

Any denier-relatives (or anyone else) that you may have out there might take time to come around – in fact, they almost certainly will – but over time, as climate change worsens, they will have no choice but to face the facts and act. Unity can be found here, and the earlier you start, the easier it will be later on.

Individual actions – like talking to deniers, eating less meat, taking fewer flights, or even just turning the heating off for a bit – might seem futile. However, personal experiences can help you focus on change, give you tangible moments that can be discussed with deniers, and give direction rather than aimlessly worrying about how enormous and unfathomable the problem of climate change is.

CHAPTER TWO

CLIMATE HISTORY

THE EARTH WAS FINE WITHOUT US

One point that climate deniers like to bang on about is the fact that the Earth's climate changes naturally. To be fair, they're not wrong. Saying our planet is not old or ever-changing is like saying that Beyoncé is overrated. She's not, she's a verified queen, and the Earth has gone through more permutations than there were different members of the Sugababes. However, what's happening now is unlike *any* of the other climate-changing events or cycles that have occurred in the *whole* of our planet's history. Literally all of it – like, billions and billions of years.

So, in order to understand how drastic the damage that humans have done to the Earth is, we must first get a grip on what has happened to this planet before, and thus cast our gaze back to the time before time itself... or a long way back anyway.

The headline is that climate swings occur approximately every 100,000 years for various complex reasons but – and this is the key – this in no way explains the crazily rapid levels of heating that we are seeing today. In fact, based on past trends in data, we are supposed to be in a period of *cooling*, not heating.

And throughout this section, don't forget that climate and weather are not the same thing. Weather is something that changes in extremes daily. Climate is long-term weather patterns. Got it? Good! Okay, let's dive in.

EARTH HISTORY

IF DEEP TIME WAS A CLOCK

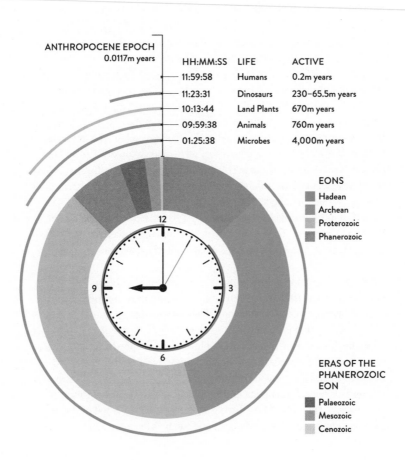

ANTHROPOCENE EPOCH
0.0117m years

HH:MM:SS	LIFE	ACTIVE
11:59:58	Humans	0.2m years
11:23:31	Dinosaurs	230–65.5m years
10:13:44	Land Plants	670m years
09:59:38	Animals	760m years
01:25:38	Microbes	4,000m years

EONS
- Hadean
- Archean
- Proterozoic
- Phanerozoic

ERAS OF THE PHANEROZOIC EON
- Palaeozoic
- Mesozoic
- Cenozoic

Let's talk about deep time, and no we don't mean telling your best mate how much you love them after too many beers. We're talking about the 4.5 billion years that this planet existed before humans even popped up. Imagine the lifespan of the Earth in a year. The Earth's born on midnight of January the 1st (happy birthday, Earth!). It's just chilling as a fairly barren rock until February 25th when the first signs of life, as single-celled organisms, appear. In March, photosynthesis (the process by which plants turn light into food and biology 101) makes its debut. In August, multicellular organisms rock up, and by September they are getting it onnnn (first instances of sexual reproduction, nice). Fungi, fish, land plants and insects all show up to the party in November, with amphibians and reptiles taking a bow in early December. Mammals, birds and flowers crop up between the 13th and 20th of December. The dinosaurs go extinct on Christmas Day (guess they were on the naughty list). And now it's New Year's Eve again, and at 11.30am hominids learn to walk. They're just bopping about until the first recorded *Homo sapiens* (us) appear at 23.36 in the evening. Agriculture appears at 23.59 – that's one minute to midnight. The industrial revolution takes place at exactly 23:59:58, two second before the end of the year. Yep, you read that right. The human species has done **ALL THIS DAMAGE** in two tiny seconds of annualised Earth time.

What was happening before humans screwed it up? To begin with, 4.5 billion years ago, there wasn't an atmosphere on Earth to speak of, but once the crust had cooled volcanoes began spewing vast quantities of gas, mainly carbon dioxide and water vapour, along with lots of ammonia and methane. It was just a planet covered in volcanoes spewing up minerals, steam and hot molten lava, and being pounded by asteroids during the late heavy bombardment 3.8 billion years ago. It was probably as fun as it sounds. Eventually, comets and geophysical events caused water to accumulate. The atmosphere then was warmer than it is today, due to the increased presence of greenhouse gases such as ammonia and methane (for more on greenhouse gases, see page 90). Over time, though, the planet slowly started to cool, creating a condensing atmosphere where rain and water cycles could occur. Those water cycles, erosion and plate tectonics created landmasses distinct from the water. Geology, atmospheric chemistry and physics all interacted to start creating the planet that we know, and soon biology got in on the mix too, with the first signs of life.

"BUT WHILE WE RECOGNISE THE OCCURRENCE OF THESE NATURAL, CYCLICAL ENVIRONMENTAL TRENDS, WE CAN'T SAY WITH ASSURANCE THAT MAN'S ACTIVITIES CAUSE WEATHER CHANGES."

Republican Governor Sarah Palin
not really grasping the severity of the situation...

CLIMATE HISTORY

So we've got land, we've got water, and we've got gas. Then about 3.5 billion years ago bacterial life evolved. But the Earth's climate doesn't change for no reason, so what happened? Well, over billions of years, a hell of a lot happened, obviously. But one major change occurred around 2.4 billion years ago: the Great Oxidation Event. At this stage, life on Earth consisted of anaerobic microbial life forms. These little fellas did not require oxygen to respire and gave off hydrogen sulphide, which smells like rotten eggs, ick. Basically they're gross, and it's scientifically proven that your boss evolved from them. Just kidding: they pretty much all died out because oxygen was toxic to them. But where did this oxygen come from?

It seems likely that photosynthesising (ie. oxygen-producing) prokaryotic organisms evolved around this time, and as the population of these organisms grew, the atmosphere contained more oxygen, and thus become increasingly toxic to the poor old microbial life forms. Slowly but surely, the oxygen-producers took over the whole world, changing the climate irrevocably in the process, and causing the deaths of pretty much everything else living on Earth at the time. Sound familiar?

Fortunately, in this case we were rooting for Team O2, and once the anaerobic bacteria started to die off, cyanobacteria took over, excreting oxygen just like plants do. These friendly little guys paved the way for algae and aerobic microbes to take over, eventually allowing for the evolution of multicellular organisms much like frogs, horses, owls, snails, dogs and ourselves. This is evidence that even tiny single-celled life forms can dramatically change the atmosphere, climate and the course of life history itself. So imagine what humans could do!

The Earth and its climate has always fluctuated between two different extremes called greenhouse and icehouse climates. Greenhouse climates are more watery, and icehouse climates are... more icy. The Earth has been a greenhouse for around 85% of its history, and greenhouse climates are hot as hell. There is more CO_2 in the air, more water on the planet (in the air, too) and very little (if any) ice at the poles. The world was as hot and steamy as Swedish sauna. Icehouse climates on the other hand were, you guessed it, chilly. They had more molecular oxygen in the air, and are defined by having ice sheets at both poles. We are actually in an icehouse state right now and have been for the past 34 million years – but we might not be for much longer, if we aren't careful.

The ancient drawn-out cycles between icehouse and greenhouse climates have churned along for hundreds of millions of years and depend on the relative concentration of CO_2 and O_2 in the atmosphere. Flipping between states takes millions of years, allowing life plenty of time to adapt to the new climatic conditions, survive – and even thrive.

Glacial periods are different and are thought to be more to do with variances in solar radiation. During these times, the Earth gets cooler and, when it coincides with an icehouse climate, an ice age. Over the past million years, about a dozen major glaciations have happened, the largest of which peaked 650,000 years ago and lasted for 50,000 years. The most recent glaciation period, often known simply as *the* Ice Age (excellent movie), reached its peak around 18,000 years ago before giving way to the interglacial Holocene epoch (our one) about 11,700 years ago.

Today's changes are different because they are so sudden! If we were to change from an icehouse to a greenhouse world in a few decades, it would cause widespread extinction. The Earth's climate doesn't change for no reason, so what caused the previous flips? TBC after this icy information...

ICE CORES

How do we know about stuff that happened literally billions of years ago? Well there's two main ways scientists can examine our deep history: fossils and ice cores.

By drilling down into ancient ice, like in the Arctic, we can observe air trapped by snowfall millions of years ago that, over time, has been compressed into ice. The composition of this air contains undeniable physical and chemical evidence that reveals climate trends in the past. Think about that... actual ancient air is being measured and tested. How *cool* is that?!

Ice cores go back hundreds of thousands of years – way before *Homo sapiens* existed. The oldest found was a ridiculous 2.7 million years old, but the oldest continuous records – allowing us to build up an accurate picture over time – go back 800,00 years. The very real and quantifiable bubbles of air that are trapped within them tell us that the rate of change that is happening to the climate today is completely out of character with what has gone before.

How can we measure this? Here's the scientific breakdown. Ice cores can tell us about precipitation rates through the thickness of the annual layers. They can tell us how warm summer temperatures were by showing us the thickness of bubble-free melt layers. And we can also use isotope ratios (in this case usually oxygen-16, oxygen-18 and carbon-14) to compare the presence of different types of oxygen and carbon in past atmospheres. The balance of these ratios shows what temperature the snow was when it first formed. How? Because water molecules containing heavier ^{18}O isotopes don't evaporate as easily as water molecules with the lighter ^{16}O isotope. An increased presence of

^{18}O indicates increasing temperatures, because more thermal energy would have been needed to evaporate these heavier isotopes.

There's a ton more that we can do with ice cores. For example, scientists can also see how much nitrous oxide and methane (for more on methane, see page 92) there used to be in the atmosphere, and that provides its own wealth of information. Crucially, though, ice cores are just as useful for looking at our recent past as our ancient past. Tracking our CO_2 emissions for the past couple of hundred years does not make for pretty reading, and even less so in the past 10 years. In that time, CO_2 has increased by 20 parts per million by volume. Not that much, you say? 20 in a million – that's a drop in the ocean!

Well, from all of our ice-core analysis, we know that the fastest natural rate of increase that the past million years or so has seen was when the Earth was rapidly heating and coming out of its last ice age. At that time it was increasing at **20 parts per million** by volume in a whopping 1,000 years! Much like a rapper's personalized swag, our carbon consumption is literally written in ice.

ICE CORES

0m

11,500 years ago:
End of last ice age.

21,000 years ago:
Peak of last ice age.

420,000 years ago:
Oxygen depletion in the Earth's oceans causes major mass extinction of marine life.

450,000 years ago:
"Stage 11" – a period when the Earth's orbit was similar to how it is now.

780,000 years ago:
Reversal of the Earth's magnetic field.

3200m

1 million years ago

ICE, ICE BABY

Let's get back to the dawn of time and the Great Oxidation Event. The cyanobacteria both absorbed CO_2 and excreted O_2, triggering a long cooling period and the first icehouse climate and mass extinction event. This lasted 300 million years until, for reasons unknown, the Earth snapped out of it and reverted to a greenhouse climate.

Then, around 715 million years ago, the Earth flipped again and experienced its second icehouse state and the most severe ice age of all time, creating what scientists literally call Snowball Earth (you've got to admit, the scientists who get to name these things know what they are doing!). This deep-freeze event left the Earth pretty much completely covered in ice for 120 million years. The average temperature of the equators at this time was similar to what we observe today in Antarctica. In the 1990s, geologists even found evidence of actual glaciers in the tropics. Pretty cool, huh? Freezing, even.

Fortunately, a positive feedback loop created by the hot duo of belching volcanoes and methane-farting microbes created enough of a greenhouse gas layer to thaw that snowball. That brings us up to the past 590 million years, during which the earth and its climate has fluctuated between the two different states fairly regularly.

About 55 million years ago the planet reached peak heat at a time that scientists call the thermal maximum, caused by increased carbon in the atmosphere. At this time, the Earth was even hotter than Nigella Lawson dishing up a Christmas dinner. H-O-T! Tropical forests covered the Earth, even as far as to the North and South Pole. Fossils of giant turtles – who bloody love warm water – have been found in both the Arctic and Antarctic. This thermal maximum global warming event sent temperatures rising by between 5 and 8°C over a mere 20,000 years. The long-term effects of this balmy world were enormous. It enabled fish, birds and animals including marsupials and hominids to evolve. What's more, flowering plants and pollinators evolved and thrived, as did the azolla fern.

You've probably never heard of it, but this one plant species did more to change the course of life on Earth than pretty much any species until... well... us. According to fossil evidence, the aquatic azolla fern was incredibly abundant 49 million years ago, especially in the Arctic. When the ferns died, they would sink to the bottom of the sea floor, taking all the CO_2 that they contained with them. Because of a freak series of conditions, these ferns did not decompose, and so the azolla – and the CO_2 they contained – was buried in the seabed for good.

Unbelievably, this innocent-looking fern was responsible for removing 80% of the CO_2 in the atmosphere over a period of just 800,000 years. Prior to this, the Earth was sweltering in its greenhouse state, and life on Earth was enjoying it like it was a naked Scandinavian in a hot tub. But, thanks to the azolla, the Earth swiftly entered an icehouse phase. The ice ages that resulted saw the Earth freeze over and for life on Earth, things were tough – imagine the Scandinavian stepping out of the hot tub and right into an Arctic winter.

So, yes, the climate has changed dramatically before, driven mainly by the changes in the atmospheric ratios of CO_2 and O_2. And, yes, other natural events and species have been responsible. But it's also important to remember that those changes have occurred over thousands of years at the very least, and even then they have been responsible for HUGE swathes of extinction. If those tiny wee plants were capable of having a large-scale global effect on the climate with relative ease, think what humans could do! And we are shaping up to cause the same sort of damage in just hundreds of years, giving life on Earth absolutely no chance to adapt and survive.

Seeing as we still (for the moment) have ice at the poles, we are still in an icehouse period of cooling... but we have started creating chaos, of course, and unless we change our actions soon, we won't be for long.

GLOBAL TEMPERATURE ESTIMATE (MILLIONS OF YEARS)

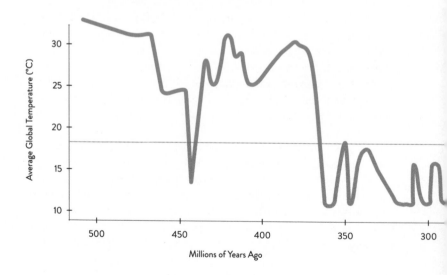

JURASSIC PARK: THE PREQUEL

Geological and fossil evidence has led experts to strongly suspect that 66 million years ago a huge asteroid impacted Earth. Not only did this cause shockwaves and tsunamis that would have killed everything in the near vicinity, but it also sprayed hot debris up into the atmosphere, cloaking the Earth with masses of dust. The climate changed so rapidly that 75% of all species became extinct, including the large dinosaurs, allowing the then-small mammals to take over and dominate the world.

The rapidity of the changes that humans are making in the climate is probably only matched by this once in a 100 million year event, and that caused **75% of ALL SPECIES** to die. Is this a price that we are willing to pay?

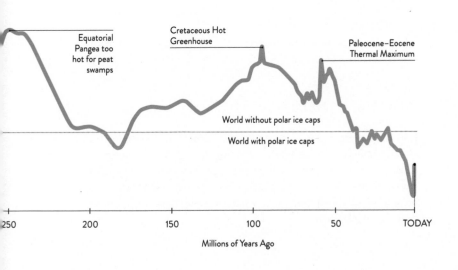

HUMANS MESS IT UP

In the past 100 years, temperatures have risen so dramatically that the natural gradual cooling over the past 6,000 years has been completely cancelled out. According to Global Circulation Modelling (see page 18) we should be in the middle of a cooling period lasting thousands of years more, but this is clearly not what we see happening. Human activity is the cause of this hot mess, and while the planet will probably survive one way or another, what will become of humans?

It's easy to think that humanity has progressed: let's be real, you can literally order a pizza, a yoga mat and a pair of jeans to your door without even getting out of bed. That's progress, right? Sure it is, for a handful of privileged people with Domino's Loyalty Cards, but progress is not even and climate change affects those at the bottom of the ladder the worst.

Climate change didn't happen overnight, so what changed? Here's a whirlwind history of humanity and the background to how we got here:

HUNTER-GATHERERS (1.8 MILLION YEARS AGO)

Humanity originated in Africa and spread from there across the world. Hunter-gatherers were already impacting the climate even at this early stage, as they killed off the likes of woolly mammoths and saber tooth tigers, along with many other megafauna. The extinction of these giant creatures had a big impact on ecosystems, leading to other animals adapting to take their place.

AGRICULTURAL REVOLUTION
(13,000 YEARS AGO)

Like an overfed cat spotting a sunbeam, humans settled. Enough change
(cultural and technological) had occurred for agricultural practices
to spread out and support increasing populations across the world.
Communities who chose to settle into agriculture pushed hunter-
gatherer communities onto less fertile land, forcing more long-distance
migration. During this time the first alphabets, numerical and writing
systems were also developed.

MERCANTILE CAPITALISM
(600 YEARS AGO)

This is the prequel to modern-day capitalism no one asked for,
but you're still getting. Mercantile capitalism brought about the
accumulation of enormous amounts of resources by the empires of
Europe, and the result was exploitation and slavery. While previous
civilisations (Ancient Egypt, Ancient Greece, the Roman Empire, the
Mongol Empire, the list goes on...) had robbed and enslaved *only* their
neighbours (good for them), mercantile capitalism was the first era to
include an intense, rapid and vast spreading of terrestrial plants and
animals across the whole world, causing the permanent disruption of
both human communities and entire ecosystems.

Take, for example, the Americas. As well as bringing European cattle, horses
and earthworms to a new continent, the "settlers" also brought diseases
that wiped out countless Indigenous peoples (just to rub it in, survivors
were subject to genocidal mass murder and cultural erasure at the hands
of European missionaries). If the near extinction of a whole people wasn't

enough, over 12 million people were kidnapped and trafficked from Africa to the Americas, forced into slavery.

Not only were plantations despicable for the amount of human suffering they caused, plantation practises were disastrous for ecosystem diversity and health. Monocultures such as coffee, cotton and sugar made European superpowers, aristocracies, and colonisers extremely wealthy, but absolutely destroyed the natural habitats of the Americas in a matter of decades. The subsequent nihilistic economic boom served only a minority of rich heirs who profited from a system of misery and environmental devastation, a rotten legacy that we still unfortunately live with today.

INDUSTRIAL REVOLUTION (300 YEARS AGO)

Ah, the Industrial Revolution. The golden age of smog, child labour and general misery. As with mercantile capitalism, the Industrial Revolution made massive profits for the 1% while simultaneously increasing pollution, urbanisation, coal burning, cotton farming, forced labour and inhumane working conditions.

If you've read any Charles Dickens (or watched *The Muppets Christmas Carol*) you'll know that cities weren't exactly nice places to be in this era. Not only were people expected to work from the age of four until they dropped dead at the ripe old age of 27, but the Industrial Revolution once more accelerated the increasing ecological disruption around the world.

Thinkers of the time attributed pollution and ecological disturbance to overpopulation, nicely excusing those at the top. However, many scholars of Victorian history today point out that it was not overpopulation but rather the unjust economic system those "overpopulated" masses toiled under. Incidentally, a by-product of this overpopulation myth stoked the fires of eugenics and fascism, which were used to justify atrocities such as the Holocaust. Just lovely.

THE GREAT ACCELERATION (70 YEARS AGO)

After World War II, the Western economy became driven by consumer capitalism. Known as the Great Acceleration, the use of petrochemicals and dependency on fossil fuels increased rapidly, as did the use of synthetic chemicals (plastics, anyone?!), causing the climate issues we are all too aware of, and that will be explored in more detail later in this book.

From the 1980s onwards, people became more aware of the impact of industry on the planet, as well the negative impacts of a rampant free-market capitalism. What's more, the advantages of the Great Acceleration were by no means equitably shared. For example, while many medical and technological inventions currently make human lives longer and easier, the benefits of technology and medicine are not even remotely equally accessible.

Which brings us up to the current time. All these factors and the weight of history have condensed into the boiling point we are about to reach. We must consider what lessons can be learned from history – not only where we went wrong, but just as importantly, where developments have been seen to improve people's lives. Inequality and environmental damage go hand in hand. Whatever happens next in our history, our collective aim must be for everybody to gain from technological and economic developments, for a better human race and for a better planet.

"IT HAS BEEN PROPOSED THAT ECOCIDE, THE ENVIRONMENTAL EQUIVALENT OF GENOCIDE, BECOMES THE 5TH INTERNATIONAL CRIME AGAINST PEACE ALONGSIDE GENOCIDE, CRIMES AGAINST HUMANITY, CRIMES OF AGGRESSION AND WAR CRIMES. UNDER THE PROPOSED NEW LAW, HEADS OF STATES AND DIRECTORS OF CORPORATIONS WILL BE REQUIRED TO TAKE INDIVIDUAL AND PERSONAL RESPONSIBILITY FOR THEIR ACTIONS."

Polly Higgins, *Eradicating Ecocide*, 2011

THE "HOCKEY STICK" GRAPH

We've already seen that periods of both warming and cooling are natural to the planet, so why all the fuss? Sure, change is natural, but the **rate** at which we're seeing the Earth's temperature change is unprecedented, unnatural, unstable and unsustainable.

There is a measurable connection between the amount of CO_2 in the atmosphere and the temperature of the global climate. Since the 1850s (read: the Industrial Revolution), we can chart an incredibly (not in a good way) rapid rise in CO_2 and in temperature.

Popularised by the climatologist Jerry Mahlman, the term "hockey stick graph" describes the pattern shown by research concerning past temperature trends. This graph shows the global average temperature record over the past couple of thousand years based on evidence from environmental resources (such as ice cores, see page 40). The graphs, now widely used by climate scientists, present data that can only be interpreted as a gradual cooling trend that changes into a rapid warming from the 1850s onwards. The evidence is absolutely damning – humans are screwing the planet up.

HOCKEY STICK GRAPH

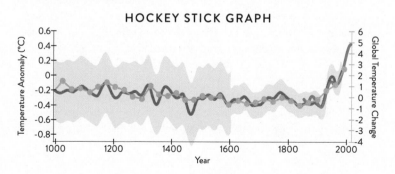

MASS EXTINCTIONS

Did you know that 99% of all species that have ever existed are now extinct? Now you do!

Although mass extinctions are hugely disruptive events, they create space on the planet for new forms of life to emerge and thrive. For example, the extinction of the dinosaurs led to the evolution of most modern mammals. It also led to at least one child in your life being annoyingly into dinosaurs.

The dinosaurs getting wiped out was super fast – that kinda happens with asteroids – but nearly every other species extinction happened over a course of hundreds or even thousands of years. The rate of extinction we see today is **extreme**, with some species being wiped out a matter of decades after having perfectly healthy, happy populations.

The fossil record tells us not only *what* kinds of creatures once roamed this Earth, but also –using carbon dating – we can tell pretty accurately *when* they lived. Armed with this information, scientists have identified six major extinction periods:

• Ordovician-Silurian extinction (444 million years ago)
In the second-worst mass extinction known to science, this event killed an estimated 85% of all species. The formation of glaciers locked up huge amounts of water, causing sea levels to drop and taking its toll on pretty much all marine organisms.

• Late Devonian extinction (383-359 million years ago)
This was a sea change, pun intended. Oxygen in the sea dropped and 75% of sea creatures were made extinct over a 20-million-year period.

• Permian-Triassic extinction (252 million years ago)

You thought the dinosaur extinction was going to be the worst one? Think again! Also known as the "Great Dying" this was the single worst event life on Earth. In a (short) space of 60,000 years, 96% of all marine species and about three-quarters of land species died. Even the bugs couldn't escape (although I bet the cockroach would have given it a good go). Scientists believe that a massive volcanic explosion probably caused all this.

• Triassic-Jurassic extinction (201 million years ago)

It took a while for the Earth to recover, but when it did, species on both the land and in the sea diversified quickly, leading to a group of creatures called archosaurs. These were early ancestors to birds, crocodiles, dinosaurs and your old history teacher. All was going well until – you guessed it – another bloody volcanic explosion! This time, 80% of land and sea creatures were wiped out. The only creatures left were reptiles called....

• Cretaceous-Paleogene extinction (66 million years ago)

.... Dinosaurs! Ah yes, the dinosaurs, they had a jolly old time romping about the place, eating each other, not being able to scratch their own backs with their teeny-tiny little Tyrannosaurus Rex claws. You already know the drill on this one; meteorite, boom, bye-bye dino-babes.

• Biodiversity crisis (Today!!)

Scientists have warned that we could face the next period of mass extinction in as little as **240 to 540 years**. Remember that most naturally occurring extinctions happen over hundreds of thousands of years, or about as long as it takes for you to get a text back on Tinder. Estimates show that over one million species are currently facing extinction. Threats to life on this planet no longer come from random meteorites or volcanoes but from the likes of deforestation, overfishing, the spread of invasive species and diseases, as well as pollution and human-caused climate change.

PLAYING GOD: GEOENGINEERING

The Earth is constantly changing and evolving on its own lengthy time frame. Whether it's ice ages or volcanoes, the Earth is still powerful beyond human capability. So, maybe it's not the best idea to mess with it too much?

Some bright sparks have embraced geoengineering as a solution to climate change. Sometimes called climate engineering or climate interventions, geoengineering involves technological large-scale interventions intended to purposefully transform the climate system. The objective is usually to mitigate the adverse effects of climate change.

Almost all experts and major scientific reports advise **against** banking on geoengineering as the solution to global warming. Simply, the whole concept involves such large uncertainties concerning effectiveness, side effects and unforeseen consequences that it is seen as too risky to try. There was even an awful 2017 Hollywood movie about it, *Geostorm*. Do yourself a favour and stream it, because it's hilariously bad. Full Gerald-Butler-out-running-lava bad.

There is substantial agreement between scientists that geoengineering cannot and should not be a substitute for climate change mitigation. In the future, some methods and interventions might be useful or even necessary (such as afforestation), but they should always accompany sharp cuts in greenhouse emissions instead of simply plastering over the problem and allowing Big Oil to keep pumping. Even if, as a species, we did want to bet the house on it, the jury is still very much out on whether or not geoengineering methods are safe, reliable or ecologically sound.

SOLAR RADIATION MANAGEMENT

Solar radiation management (SRM) is an idea that aims to offset the greenhouse effect by using reflective surfaces and particles to mimic the effect of ice sheets that would naturally cause less solar radiation to be absorbed by the Earth. In layperson's terms, it's giant mirrors.

These mirrors would bounce the Sun's rays back out into the atmosphere and cool the ground temperature. Of all the geoengineering suggestions out there, this is one most likely to *possibly* work. Simulations have shown that solar radiation management might be able to stabilise global temperature rise. Whether or not this will actually work on a large enough scale is not yet known. In the best medical advertisement voice we can muster: possible side-effects include disrupted weather patterns, such as changes in precipitation that would cause its own side-effects that we can't even begin to predict.

....

"THE EARTH IS NOT OUR PRISONER, OUR PATIENT, OUR MACHINE, OR, INDEED OUR MONSTER. IT IS OUR ENTIRE WORLD. AND THE SOLUTION TO GLOBAL WARMING IS NOT TO FIX THE WORLD, IT IS TO FIX OURSELVES."

Naomi Klein, *This Changes Everything*, 2014

....

AFFORESTATION

Planting more trees is always a good thing, right? If done correctly, absolutely! But this plan also requires careful consideration.

Firstly, there is a difference between afforestation and reforestation. Afforestation is introducing trees and tree seedlings to areas that would naturally not be forested, while reforestation is replanting in an area where forests used to be. However easy it might seem and good it might feel to stick a few trees in the ground and go on your merry way, there are far more things to consider. For example, planting the wrong plants in the wrong places could disrupt natural ecosystems just as much as climate change would otherwise do. It could also disturb local insect animal populations, or spread other non-native and harmful species that came along for the ride with the trees.

It's possible that there is an even better option out there, especially for those who love the ocean spray in their face: the use of seaweed farms to remove CO_2 from the atmosphere. This seaweed could then also be used as a food or as a biofuel source. What's more, growing the seaweed could also help counter ocean acidification by absorbing CO_2 from the water. Win-win-win... what's not to like?

And if the real deal isn't enough, artificial plants and trees have been proposed that could directly capture CO_2 and either store it as a gas indefinitely or – even better –use it as a biofuel. One prototype in Iceland has used the captured CO_2 to make artificial basalt (a stone) to form carbon-containing minerals.

PROBLEMS WITH GEOENGINEERING

These are some of the more sane ideas out there – there are some truly bananas ones that are as fun to read as they are unlikely to ever happen, like the people who want to spray paint most of the Earth to make it more reflective! The giant mirror ideas was bad enough...

Contentious already because of the practicality and usefulness of each ides, geoengineering is also controversial because it suggests that we can just employ a "technofix" rather than change our own ridiculous behaviour. The suggestion is that we can just keep using and abusing the Earth as we have done over the past 250 years and these ideas will give us a quick and painless way out.

Given that there is so little time and political will (sometimes even flat-out refusal) to do something about climate change at this point in time, conducting risky global-scale alterations to our planet using untested technology is dangerously attractive to the political classes. Why would you bother cutting carbon emissions if you can invent something that sucks them out of the atmosphere altogether?! However, this is a precarious path to tread, and it displaces the responsibility from us to actually change our damaging systems and lifestyles.

Overall, geoengineering solutions cannot be considered serious climate saviours, as much as tech tycoons like Elon Musk might want to. There is too much unknown about how they would affect even the most basic of cycles, like the weather systems on Earth. Is it worth the risk? Well if you think you can outrun lava or avoid a frozen aeroplane falling out of the sky like in *Geostorm* (spoiler alert – although surely you've watched it by now?) then maybe, but until then we need **real solutions** to cutting emissions and cleaning up the turd mountain of damage we've done to the planet.

CHAPTER THREE

BIODIVERSITY

WHAT IS ECOLOGY?
FT. FRED THE ANT

Stop reading for a moment and get your nearest music-playing device. Got it? Good. Now play "The Circle of Life" from *The Lion King*. In its simplest form, that's what we're about to talk about: the circle of life.

In ecology, everything – and we mean everything – is connected. Every creature has its place in the inter-connected web of its habitat, or home. Even the very word "ecology" originally comes from the Greek word *oikos* meaning home. The lives of *all* organisms – including humans – are deeply connected to their environment, and ecology explores these interactions between creatures, plants, humans and their environment.

In this chapter, we will be briefly exploring ecological principles that are important to understand, especially in terms of the Earth's climate. What it boils down to, though, is that everything is interconnected within a biosphere. The health or abundance of one species will inevitably affect the others. No (hu)man is an island, and that applies to all life on planet Earth!

While ecosystems contain space for natural diversification of their inhabitants, they are also highly organised. Take a common ant, for example. Let's call this ant Fred (hi, Fred!). Fred is just one ant but he lives in a colony. All the ants in this colony are called a population. Fred's colony lives in the ground, in an extraordinary nest made by Fred and his pals moving earth around. Fred's population might feed off plants or seeds or other food, or even fungi. Poor Fred might get eaten himself by another insect, or perhaps a bird if he's not careful. Fred is part of a community that he is both sustained by and contributes to.

Multiple little communities that co-exist and interact in an environment are called an ecosystem. For example, the bird that eats Fred could be eaten by an eagle, but that eagle can't survive on little birds alone so it also eats rabbits. Fred doesn't hang out with rabbits – at least, not that he'll admit to in front of his other ant friends – but they co-exist in the ecosystem nonetheless. They both share the same soil and plants, and they both interact with the same other species.

But think even grander for a moment. In any given area, there are thousands of species of plants and animals, all interacting in different ways. Fred the ant is... well... just an ant. His ecosystem is one part of a network of ecosystems that collectively form a biome. Common biomes include forests, deserts, riverbanks etc. Combining all the various biomes together forms the Earth's biosphere.

As you can imagine, remove any single species, disturb physical habitats or tamper in even a small way with this delicate natural order, and you risk bringing an ecosystem down like a pyramid of cards. One ecosystem will, inevitably, affect the others that interact with it. Abundance – the number of organisms in a population – and the rates of birth and death in a population, affect this delicate web as well. So, remove one species, and you could inadvertently cause a cascade of events that affects everything else. Even when it's something as small and seemingly as inconsequential as an ant... no offence, Fred – you're our favourite.

Because of all of this, matters of ecology are life and death situations. Literally. The food web and how energy gets passed from one creature to another is a matter of life or death, not only to animals, but to humans as well. Zoom out again: energy comes from the sun, right? The movement of energy (such as sugars and starches) and nutrients (minerals, salts and elements such as carbon, nitrogen, phosphorus and oxygen) through an ecosystem maps out very intricate paths, called a food web. All the energy an organism needs to live comes from chemical reactions that take place within them and depends on the nutrients they consume. Nutrients cycle through ecosystems through feeding, defecation and excretion, death and decay (it's really fun to think about). As we said: the circle of life.

You can see that all living organisms in an ecosystem are codependent and live in a tight interaction with each other and their environment. Importantly, all organisms can and do adapt their behaviour to optimise their chances of survival. Think of giraffe's long necks, developed over time to reach the leaves that shorter rivals couldn't get to. Do you think those necks just grew overnight? Or that suddenly a giraffe was born with a neck eight times as long as its mum's (think of that poor, poor mother giraffe!)? No, of course not. It takes a long, long time. The rate of change in diversity and adaptation is an important factor in the health of an ecosystem. It can be measured and predicted by scientists, and this evolution and adaptation is a natural part of the world around us.

But natural adaptation is *not* what is happening here, and *won't* magically save the day. Human populations have transformed ecosystems and whole biomes so drastically that the natural rate of evolution can't keep up. Remove Fred and his pals from an ecosystem over a millennia, and the birds that feed on him will have a chance to find other food sources. Population might fluctuate, and they will have to evolve and adapt, but the chances are that they will survive. Remove Fred overnight – or in a matter of years – and they have a much greater struggle on their hands. Then, if there are fewer birds because they have less food, what happens to the eagles? Maybe they have to eat more rabbits. And so on, and so on, and so on. Even minor human-induced changes to an ecosystem can cause dramatic knock-on effects that can accumulate at a break-neck pace. And just in case you think, "I don't really care about Fred, or his *Animal Farm* pals..." firstly, what kind of monster are you? What's Fred done to you? Secondly, humans also need stable ecosystems, not only for our food, but for raw materials (wood, for example), or even for our own mental health.

You would think that, because of the importance of ecosystems, governments and decision-makers would take into account whatever changes might occur when planning new policies. It would be very nice to think that, anyway. But unbelievably, biodiversity is taken for granted in almost all economic calculations as something that will stay static and unchanging, even as we raze forests and cull wildlife. It's bananas. It's nuts. It's cuckoo.

Biodiversity and ecosystems provide regulating, provisioning, supporting and cultural services that contribute to everything humans do. If we are not going to wreck that for Fred, for Eddie (we've decided to call the eagle Eddie now) and for humans, we need to start treating them as the invaluable asset that they are.

"THE TWO BIGGEST HUMAN THREATS TO WILDLIFE IN THE LAST CENTURY HAVE BEEN A) COMMUNISTS AND B) ENVIRONMENTALISTS."

Dr Patrick Moore, a high-profile climate-change denier and Fox News favourite who (randomly and incorrectly) claims to be a founding member of Greenpeace.

EXTINCTIONS

While extinctions do naturally occur, the rate of extinctions that we are seeing today is like nothing that has been recorded in recent millennia. That's even taking into account extinction events like humans pitching up on a new island, remarking "this is lovely," before systematically clubbing every last dodo on the island to death because we were "bored of chicken".

The UN has found that around 1 million plant and animal species are currently at risk of extinction. Quite simply, we are living through the world's sixth mass extinction event with the rate at which species go extinct up by somewhere between **100 and 1000 times** what it normally is.

Fortunately, help is at hand. The issue of biodiversity depletion is so serious that the UN formed the Convention of Biodiversity in 1992, which meets every three years to allow nations to discuss the most pressing issues and talk solutions. To show just how seriously they were all taking it, in 2010 they set 20 biodiversity targets to be achieved by 2020. Job done, we've probably nailed it by now, right?

Of course not! The UN found that between 2010 and 2020 governments have failed to meet *any* of the 20 internationally agreed biodiversity targets. Among the 60 sub-targets assessed, 7 have been achieved, 38 have shown progress and 13 elements have shown no progress whatsoever. Progress remains unknown for two. But don't worry: they are meeting again in 2021 to set some new targets, so we will be fine.

Stupidly, biodiversity is simply not currently part of mainstream thinking in terms of policy and law. Funding shortfalls and harmful subsidies to oil, logging, farming and fishing industries have made even the limited goals set so far unachievable. Habitat destruction remains "worryingly high", according to the report – quite the understatement. Another major scientific report published in 2019 by the Intergovernmental Science-Policy Platform on Biodiversity and Ecosystem Services (IPBES) has also warned that mass extinction is accelerating, and ecosystems across the world are deteriorating at unprecedented rates.

The alarms are all going off. And yeah, biodiversity is in a whole world of trouble! Unless we find ways to break this toxic pattern at the highest governmental levels, we can say goodbye to millions of animal and plant species in the next few decades.

TROPHIC CASCADES

No, trophic cascades are not the long-anticipated follow-up to *Tropic Thunder*. Although we can't wait for that. In an ecosystem, when a population of an organism is either introduced, or becomes extinct, or over-populous, it can change its entire balance. This process is called a trophic cascade. In ecology, the word "trophic" means anything referring to feeding and nutrition. Thus, when populations of other species are negatively affected by a trophic cascade, nutrient cycles can become destabilised, creating a domino effect of extinctions and general unpleasantness.

Much like classic drinking games, trophic cascades generally work in two different directions: top-down and bottom-up. Top-down happens when trophic levels are impacted by destruction of a predator population, i.e. carnivores normally keep hungry herbivores at a reasonable number which stops them eating all the plant life, which would cause negative effects like destabilising soil. Remove the predators, and the herbivores run amok.

Bottom-up works the opposite way: this is when grasses and plant growth are restricted for some reason, for example by habitat destruction, plant disease or human construction. If there aren't enough plants to eat at the bottom, there won't be enough nutrients entering the food web to sustain populations of everything else. Either way, the ecosystem becomes much like the losers of the afore-mentioned drinking game: unstable, vulnerable and prone to causing huge issues for everyone else!

This section has focused mainly on land animals, but trophic cascades are also happening regularly in the Earth's oceans. Top predators like bluefin tuna, trout, salmon and sharks are being overfished, causing devastating issues for the biodiversity and health of the oceans. For more on oceans, see Chapter 6.

BIG ANIMALS

Because they are at the top of any food web, and thus rely on everything around them working harmoniously for their nutrition, the species that are most at risk are the largest (and, dare we say, the cutest and best-loved by humans). Large herbivores like wild horses, elephants, beavers and wild boars function as ecosystem engineers, extensively creating and modifying habitats. They transform terrain, work the soil and create openings in dense vegetation which affects everything else from large rivers to tiny micro-organisms. Predators such as wolves, tigers and lions influence the behaviour and population of these herbivores – giving shrubs the chance to grow and forest borders to expand rather than herbivore populations growing so large that the vegetation is stripped back. As we have seen, everything in an ecosystem is connected, so the extinction of even one of these big animal species, on either the predator or herbivore side, can cause an even bigger problem.

MAN'S BEST FRIENDS?

Much of the Earth's ancient wilderness has been destroyed by human development, and most of this has been to make room for agriculture and the extensive farming of domesticated animals. Ecologists can measure which way the scales are tipping for wild animals by calculating biomass. This is the mass of living organisms (literally every living thing) that are present in an area or ecosystem. According to a 2018 study into the biomass distribution on Earth, livestock outweighs wild mammals and birds by a factor of ten. Yep, that's right, only one in ten mammals now on earth are wild! Even these are being pushed into smaller habitats.

Pigeons, rats, and domesticated animals such as cows and dogs are doing really well out of humans – but as much as we all love rats, these few successful species do not make for a biodiverse world. In fact, enormous populations of domesticated animals bring their own issues with them, whether that be contributing directly to climate change (see beef farming, page 122) or through the carrying of diseases that affect both animals and humans (think of COVID-19 and its prequels, like swine flu and bird flu).

....

"THERE ARE SOME FOUR MILLION DIFFERENT KINDS OF ANIMALS AND PLANTS IN THE WORLD. FOUR MILLION DIFFERENT SOLUTIONS TO THE PROBLEMS OF STAYING ALIVE."

The godfather of conservation himself,
Sir David Attenborough nails it.

....

EXTINCTION RATES

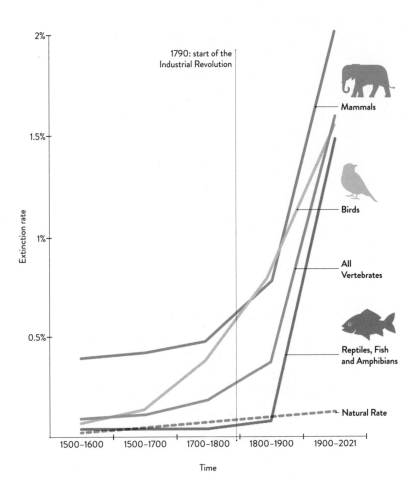

CREEPY CRAWLIES

Look, they may not be as cute and cuddly as a fluffy tiger cub or a playful dolphin, or even a rambunctious rat, but insects play a vital role in maintaining our ecosystems, not to mention life itself.

As with other animal species, insect extinction rates are currently **100 to 1000 times higher** than naturally occurring rates, and more than 40% of the world's insect species are currently in decline. One way scientists measure the health of insects is through biomass: imagine the world's largest scales, with the world's most disgusting pile of wriggling, writhing and retch-inducing insects on them. Every insect in the world, in fact. Each year for the past 30 years, that pile has decreased in weight by 2.5%. There are many causes for this, including: the use of pesticides, fungicides, agrochemicals and chemical fertilisers; diseases; habitat destruction; air and water pollution; and monoculture farming (planting large amounts of the same crops).

If you were hoping we would tell you this isn't a big deal because nobody likes bugs, unfortunately you're out of luck because we aren't gonna worm our way out of this one. Insects truly are the unsung heroes of ecosystem health. They provide essential services such as purifying water, pollinating crops, and providing food for other organisms. And it's not only the insects above ground that deserve the recognition. The ground beneath our feet is teeming with life. Insects, microbes, fungi, amphibians, reptiles, mammals and even birds nest in the ground. These underground ecosystems are perhaps even more delicate than those above-ground. Damage to them and to underground diversity can threaten the health of the soil itself, leading to soil degradation (see page 77), which is pretty disastrous for us humans.

PLANT LIFE

There's more to plants than buying a supermarket succulent and letting it slowly die in your tiny apartment. Plants are vital to all life, everywhere. Not only do they produce the oxygen that we humans need to breathe but they provide food and shelter for countless insects and animals. Trees regulate water flow, shrubs and root systems play a role in healthy soil structure, and the fungi that digest dead matter and produce rich soils live in symbiosis with these plants and their root systems. They do all this and much, much more. Basically, plants are amazing.

Yet, one in four plant species are currently threatened with extinction. Some of this is through the obvious clearing of the natural world for logging, or to clear space to build more of the aforementioned tiny apartments. Plants also bear the brunt of disruptive large-scale farming practices. And some plants are simply vulnerable to extreme weather conditions, like those brought on by climate change, and are struggling to survive because of this.

It pays to be a tree-hugger (not in cash... just in kudos), and stopping loggers chopping down forests is fantastic, there's no doubt about it. But it will take humans addressing the entirety of climate change for the challenges facing plant life to be reversed.

"IN THE LAST 40 YEARS, THE AVERAGE ANNUAL EXTINCTION RATE WAS TWO PER YEAR. NOT 2,000. NOT 200. TWO. THE MOST RECENT DECADAL DATA AVAILABLE WAS FROM 2000 TO 2009 AND ONLY FOUR EXTINCTIONS WERE REPORTED IN THAT TIME FRAME."

Gregory Wrightstone, a member of the powerful pro-fossil fuel thinktank, the Heartland Institute. These guys spent much of the 1990s arguing that second-hand smoke was absolutely fine and lobbying against smoking bans, so we know their arguments are completely based on science and have nothing to do with their funding from fossil-fuel or tobacco companies...

RESTORATION AND REINTRODUCTION: THE YELLOWSTONE WOLVES

Diverse ecosystems are stable, healthy and able to adapt to change. The reintroduction of apex predators, and other rewilding techniques might make humans feel uncomfortable, but it allows ecosystems to return to the natural balance that has taken millennia to develop.

The advantages of this go beyond returning populations to stability; with larger populations of organisms comes greater genetic diversity, and genetic variability enables adaptation. As the climate inevitably changes, the ability for ecosystems and the organisms within them to adapt will be crucial. With rewilding, it is possible to reverse the damage that extinctions cause, as in the case of the reintroduction of the grey wolf to Yellowstone Park in the USA – which was a howling success.

Wolves once stretched from Mexico to the Arctic Circle, but years of habitat destruction and hunting saw the wolf population dwindling to near nothing. In 1926, the last two grey wolves in Yellowstone were killed, and they became locally extinct. Without apex predators, the deer population exploded and they transformed the landscape by over-eating grasses, shrubs and young trees. This in turn affected insect populations, as well as soil quality and structure. Before long, the ecosystem was being depleted and destabilised due to this over-consumption of vegetation. Oh, deer.

Through the collective work of federal and state agencies, grey wolves were reintroduced to Yellowstone in 1995. They were heckin' good doggos. Through their hunting of deer and bison, the wolves almost immediately restored balance to the park. The ecosystem recovered dramatically, rebalancing populations of grazers, plant life, insect life and soil health. As of 2020, there are roughly eight different wolf packs made up of at least 94 wolves. That's 11/10 for you, ya good wolfies.

All over the world, rewilding programmes have met with huge success. Programmes to reintroduce top predators have been shown to restore ecosystems to the healthiness they once naturally had. In 2009, beavers were reintroduced to Scotland for the first time in 400 years. Other programmes that are ongoing include the Iberian lynx in Spain, bears in parts of Europe and stoats in the UK. As well as reintroducing extinct species, there is important work being done to protect critically endangered species like lions, leopards, pumas, birds of prey and tigers.

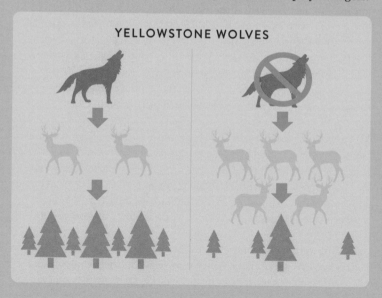

YELLOWSTONE WOLVES

HABITAT CLAP BACK

An animal without a home is like an early morning without a cup of coffee: completely lost.

The effects of deforestation can be felt practically everywhere, from Scotland to Sierra Leone, from Canada to Cameroon. Forests currently cover about 30% of the world's land, but they are disappearing at an alarming rate. According to the World Bank, between 1990 and 2016 the world lost 502,000 square miles (1.3 million square kilometres) of forest, equivalent to an area larger than South Africa.

Europe and much of East Asia has already lost almost all of its major forest cover, with extremely little wilderness remaining in continental Europe, the UK and North Africa. In the last decade equatorial regions, like South America, Sub-Saharan Africa and South-East Asia, have lost 7 million hectares of rainforest every year – roughly the size of Luxembourg being destroyed every ten days. If Luxembourg went missing, we're pretty sure people would notice... wouldn't they?

Even worse, according to the World Resources Institute, if deforestation were a country, it would be the third largest CO_2 emitter in the world. This is because trees (and the soil they hold together with their roots) absorb carbon dioxide and other heat-trapping greenhouse emissions that human activities create, so increased deforestation leads to increased levels of human-made CO_2 in the atmosphere, which is obviously really, really bad.

According to some estimates, increased tropical tree cover could provide as much as 23% of the climate mitigation needed to help control the effects of global warming over the next decade **on its own**. So not only is preserving ancient forests and planting new ones essential for biodiversity, but it is also central to the fight against rising CO_2 in the atmosphere.

The immediate effect on humans is just as dramatic. Approximately 250 million people living in forest and savannah areas depend on trees for subsistence and income – many of these are poor or living in poverty, and deforestation disproportionately affects them. Looking further afield, billions of people across the world depend on forests purely because they provide homes for the pollinators on which their crops depend.

So, forests are good for people, for animals and for the atmosphere as a whole. Why are we still insisting on chopping them down? To answer that, let's take a look at the Amazon rainforest.

THE AMAZON AND OTHER RAINFORESTS

Rainforests only cover about 2% of the total surface area of the Earth, but they are responsible for producing 20% of the world's oxygen. They are also home to 50% of all known species of plants and animals. The Amazon rainforest greatly influences regional and global water cycles through the perspiration of plants and plays a vital role in the water supply of Brazilian, Bolivian, Peruvian, Ecuadorian, Colombian, Venezuelan, Guyanan, Surinamese and French Guianan cities. Without it, it's very possible that huge swathes of South America would go very thirsty.

However, Brazil (along with countries on other parts of the world like Indonesia and Côte d'Ivoire – it is not alone in this sort of behaviour) is losing its rainforest at a record-breaking rate.

About 17% of the Amazonian rainforest has been destroyed over the past 50 years, and over the past five years the rate of this loss has increased (if you ever think that politics doesn't matter, do some research into what has happened to the Amazon since Jair Bolsonaro became president. It's not pretty). But why is this happening to such an important natural resource? Farming, grazing of livestock, mining, slash-and-burn agriculture and drilling account for more than half of all deforestation.

Soy farmers and beef ranchers clear the rainforest extremely quickly, devastating the ancient ecosystems within it. Some start forest fires that have ravaged vast swathes of rainforest, far more than they need for their crops or livestock. This uncontrolled burning is not just the quickest, but probably the worst way of clearing land: it releases enormous amounts of CO_2 into the atmosphere, and it destroys habitats more devastatingly than simple logging would do. Similarly, in Malaysia and Indonesia, forests have been felled to clear land for the production of palm oil, which is found in all kinds of products from food to shampoo.

It is easy for the western world to rant about this destruction and demand that it be stopped. However, we must recognise our own part. It is our demand that drives the market for palm oil, soy (which is the primary food source for the domestic livestock that we keep) and beef. It is also our own colonial history and the corruption that it brought that has created the systemic issues in these countries – such as rural poverty, or weak institutions that might stand up for the rainforests – that allow for this to happen. In order to stop it now, we must face up to the responsibilities of the past with actions like **providing education** and **fiscal help** to those countries that we are now asking to make sacrifices.

We already have enough land cleared globally to sustain the world's population as long as it is maintained properly. There is no need to tear down ancient habitats but we must provide incentives not to, particularly to the world's poor for whom it is a way to make short-term money.

DIRTY TALK

Sometimes you've gotta get dirty.

Not like that... we're clearly talking about soil here.

Look, we don't blame you if you don't spend all day thinking about soil. It's not the spiciest of climate change issues – unlike those sexy pandas. They have all the appeal. But soil, the very dirt we stand on, is a critical global resource, essential to wildlife and food security. Not only does fertile soil increase crop yields and provide food and financial security for countless farming communities, but it also creates a home for countless creatures. Both slimy and not-so-slimy.

The nutrients and chemicals found in soil has a huge impact on what is able to live in and grow from it. You might think that all soil is roughly the same nutritious dirt that looks so damn attractive in a muddy field, but actually it takes about 100 years for just 1cm³ of fertile soil to form and, unfortunately, a lot less time to degrade.

We don't want to harp on about it but ecosystems are a delicate balancing act, and even the tiniest of changes can cause a cascade of catastrophic knock-on effects. One such domino in that dangerous chain is soil degradation. Caused mainly by (you guessed it) deforestation, industrial farming and other industrial activity, soil degradation concerns the loss of organic matter, a decline in fertility (amount of nutrients), decline in structural condition, erosion, and a decline in the soil's ability to absorb water. Not only does the health of soil affect farming in terms of crop yields and livestock health (meaning more land would be needed to provide the same output) but soil degradation has also been shown to lead to pollution, excessive wildfires, drought and flooding. In other words, it provides a helping hand to all the best, kinda-apocalyptic "natural" disasters.

"SOIL HEALTH IS PUBLIC HEALTH."

Bill Robertson, soil scientist at the University of Arkansas, 2019 Cotton Researcher of the Year, and all-round expert in dirty talk.

So let's get down to the dirty talk already. What is currently happening to soil and how is this affecting climate change?

It is thought that about **40% of the world's agricultural land can be classified as degraded**, meaning it is more difficult to grow basic crops. Between pollution, bad farming practices and poor land and livestock management, most of the world's topsoil could be completely degraded within the next 60 years.

The good news is that there are solutions, and maintaining healthy soil doesn't have to cost us the earth. Where agriculture is not an issue, simply planting new forests can help to naturally and swiftly restore soil quality and structure. On farmland, this may not be practical, moving towards the implementation of ecologically considerate **regenerative farming practices** is key.

This sounds fancy, but the ideas are incredibly simple. Reduce dependence on chemicals such as insecticides, herbicides and pesticides, and instead of using chemical fertilisers on the soil, rotate crops with animals such as geese or pigs who will eat the weeds and poop on the ground – nature's smelly fertiliser. Planting nitrogen-fixing cover crops such as peas, beans, clovers, buckwheat and certain types of rice also recondition soil heath. Surprisingly, reducing the ploughing of farmland can also improve soil health *and* crop yield. This is because over-ploughing and tilling destroys soil structure and reduces the number of helpful microbes in it, and also affects the soil microbiota such that it becomes less permeable to water, which is extremely unhelpful for agriculture. If farmers allow water to infiltrate the soil, stop tilling and let roots and mycelia (fungi) to naturally open up the soil and give it structure, it can actually provide more effective irrigation in sown land and a better capacity for soil to retain water, rather than it just becoming muddy, boggy or arid. Finally, planting hedgerows, diverse crops and small trees alongside agricultural crops can help to maintain soil structure and safeguard from flooding.

None of this is difficult or expensive to do. However, governments need to make concerted efforts to educate and help farmers into taking these steps. They cannot expect small or family-sized farmers to be able to keep up with the latest science, or take a chance on different practices like these when their livelihoods (and, sometimes, lives) depend on next year's crops. It will only be with widespread and long-term government-led change that the threat to our soil and food supply can be averted.

SILENT SPRING AND DDT

Okay, so soil pollution may not seem all that interesting by itself, for all our attempts to sex it up with the dirty talk, but looking at the effects of harmful chemicals on the ecosystem and general environment is truly shocking.

Marine biologist Rachel Carson (1907 –1964) wrote a ground-breaking book called *Silent Spring* in 1962, following her horrifying realisation that birds were dying out across the USA in large numbers. Upon investigation, she discovered that pesticides were working their way up the food chain, damaging the environment, and killing wildlife.

The insecticide in question was DDT, a chemical that was used globally at the time. She realised that birds were ingesting DDT when eating bugs and plants, causing disruption to their hormonal function and ability to metabolise calcium, thus stopping their bodies from being able to successfully make eggshells. The majority of eggs being laid had soft, thin shells, and so many baby birds were dying before or shortly after birth, hence the accurate – if depressing – name: *Silent Spring*.

The first study of its kind, the book inspired a public outcry. Over time, public pressure against the use of DDT became so strong that governments across the world eventually banned it, including India, China, the USA and countries all across Europe. In fact, *Silent Spring* gave birth to the modern environmental movement, and eventually led to the creation of the USA's Environmental Protection Agency.

While the ban on DDT was a step in the right direction, it can be argued that it was still too little too late. Once ingested, DDT can remain in the body for decades. Even worse, DDT can also be passed into the bloodstreams of unborn offspring in pregnant mammals. Yes, all mammals... including humans. Children born decades after the ban still have traces of DDT in their bloodstream because of its cumulative effects.

Worryingly, DDT is just one example of the dangers of chemicals in the modern world. Widespread chemical use in all areas of human activity means that there are many other "DDTs" out there. Deaths in humans related just to pesticide poisoning are estimated at about 1 million every year, with many of these deaths occurring among poorer agricultural and production workers.

So, even if it's just on your allotment up the road where you're trying to grow that prize-winning two-foot long marrow (you nutter), stop using bug spray, weed killer and poisoned slug pellets PLEEEEAASSE!!!

"NO SCIENTIFIC PEER REVIEWED STUDY HAS EVER REPLICATED ANY CASE OF NEGATIVE HUMAN HEALTH IMPACTS FROM DDT."

Dr Roger Bate, an economist, free-market fanatic, proponent of DDT's use in the 21st century, downplayer of COVID-19 and exemplar of straight-up denialism. I wonder who funds AEI, the thinktank he works for?

DENIER
DENIED
DENIER

RESTORATION
AND REWILDING

Investing in and caring for biodiversity is the greatest investment we can make for our future. It is our natural infrastructure, and without it the food and water systems we depend on will fail. Biodiversity is even more important to economic development than technology itself. Without nature, and without natural systems working, we die. We don't mean to alarm you (well, we kind of do, that's what this *entire* book is about. Have you picked up on that?), but it's true.

This is not just idle talk, either. Science loves nothing more than data, and in a 2014 paper a group of scientists set about quantifying exactly how much this natural infrastructure was worth to us in monetary terms. The numbers they came back with are astounding. Ecosystem services contribute a combined value of **US\$ 125–140 trillion** per annum to the global economy. That's trillion with a T. Globally, our whole economy – as well as local and national systems of public health – depend on ecosystem and climate stability. We owe it, morally, economically, and not-being-a-dick-headedly to find solutions to habitat destruction, extinction and chemical pollution.

But finally for some good news after a bit of a downer: there are so many ways to help ecosystems! Here's a just a very small selection, but your area will certainly have established local and national projects. Look them up and pitch in, it's the least you can do for these money-makers!

GOT WOOD?

We love trees. What's not to love about them? They're tree-iffic! (Sorry...)

While we do need raw materials for wood and paper biodegradable products, how we treat logging is super important to ecosystems and the atmosphere. Logging fells countless trees per year (if you do try to count it, you'll get to 15 billion. That's roughly two for every person on Earth) and loggers – both legal and illegal – need to build roads and other damaging infrastructure to access more remote parts of forests. The way that we use trees as a resource must become more sustainable and less damaging to wildlife. Most importantly, the ancient forests still remaining and home to similarly ancient ecosystems should never be cut down. Legislation to protect them, and enforcement of that legislation, is the first and most obvious step. That means voting out people like the real-life Joker Jair Bolsonaro who just want to watch the world – and their country, and the Amazon – burn.

About 80% of all timber now comes from softwood – such as pine – which grows fast and can be farmed in a controlled way. Hardwoods like maple, oak, ash, beech, sycamore, alder and cherry are more valuable financially, but also more valuable ecologically, as they can take up to 150 years to be replaced, compared to the 25-30 years that a pine takes to reach maturity. Changing legislation and consumer habits so that the trade in hardwood is limited to only recycled materials, or a very, very limited amount of new timber each year, would be another huge step.

Importantly, protecting nature is not at odds with business or trade. Commercial forestry can sustain landscapes as well as nurture rural economic growth if done in respectful ways. The demand for biodegradable and renewable wood-based products, soft-tissue

"IF WE DO NOT HAVE MULTI-SPECIES ENVIRONMENTAL JUSTICE, IT IS NOT JUSTICE AT ALL."

Donna Haraway, American Professor Emerita in the History of Consciousness Department and Feminist Studies Department at UC Santa Cruz, and an advocate of justice for all species.

products, biodegradable packaging and wood for construction is growing, and this is a positive thing if managed in a sustainable way. For example, in many commercial forestries only 60% of the trees are harvested in order to help manage them safely.

As much as we might wish for one, there is no magical bullet to the problem of deforestation. It is a solution of logging ethically, passing laws to ensure that, and then enforcing those laws. Much (but not all) illegal logging happens in the poorer parts of the world, and that is understandable – opportunities for paying work are more limited there, and each tree is comparatively worth more. It is no use richer countries simply preaching. They need to support the poorer nations financially to stop logging, if they are serious about it.

And before you ask why it's the rich countries problem to do this (we are all interconnected, *remember*? We've just been talking about this!), need we remind you that places like Europe have already logged their forests to oblivion? That is partly why the majority of ancient forests are in less-developed countries. If we are to ask those countries to make sacrifices for all of our sakes – that rich countries themselves did not make – then it is only fair that they should contribute to make it worthwhile.

THE GREAT GREEN WALL

This ambitious proposal aims to create a 15km wide, 8000km long strip of vegetation across Africa in the Sahel, a region highly at risk of desertification. The project was first launched in 2007 to help reverse the trend of severe droughts and overgrazing that have caused desertification in the semi-arid Sahel which is one of the poorest regions in the world. The ambitious aim was to replant and restore 100 million hectares of degraded land – stopping the creep of desertification in the process – and bringing 10 million green jobs to the region. Since 2007, nearly 18 million hectares of land have been restored, two-thirds of which are in Ethiopia. This is expected to store around 300 million tonnes of CO_2 in soil and woody biomass by 2030.

However, the majority of this restored land is actually outside the de-marked Great Green Wall area. Only 4 million hectares of the 100 million hectares of the Great Green Wall's intervention zone has been restored – just 4% of the 2030 goal. The initiative has also created $90 million of revenue, with 335,000 direct and indirect jobs in agriculture and pastoral activities, water and soil conservation – around 3% of the 2030 goal. Not great when halfway through the project... but slow progress is better than none, right?

What's more, as part of the recovery programme from COVID-19, the project has been granted a new lease of life. Ministers from Burkina Faso, Chad, Djibouti, Eritrea, Ethiopia, Mali, Mauritania, Niger, Nigeria, Senegal and Sudan have agreed that improving efforts to invest in the Great Green Corridor *needs* to be part of the post Covid-19 economic recovery. They have called for the private sector to "significantly contribute" to the initiative and requested the support of international organisations such as the Green Climate Fund, the Global Environment Facility and the African Development Bank Group.

In order to reach this 2030 goal, 8.2 million hectares needed to be restored every year. In order to reach the world's goal of stopping climate change, ambitious projects like this can not be allowed to fail. And this project in particular shows environmental investment at its best: it's good for the economies of some of the world's poorest countries and good for the environment in both the short-term and the long-term. It's a win all around! If we can't make this one work, what can we do?

WHAT ELSE CAN WE DO?

There are so many more ways we can help to maintain ecosystems, from your own backyard right up to a global level. Helping at-risk animals, plants and insects really doesn't have to cost the earth.

Do yourself a favour and Google fish ladders; the videos of fishes zooming around tubes is something that everyone needs in their lives! Dams prevent some migratory fish species reaching their spawning sites in rivers and streams, so someone came up with the simple design solution of tiers of flowing water called fish ladders to help fish travel around dams.

Roads, highways, buildings and housing often interfere with the migration, foraging and hunting of many animals. From the crab migrations on Christmas Island, to snakes in Illinois, to hedgehogs in the UK and brown bears in Canada, the use of bridges or tunnels through/over/under/around human-made obstructions can reduce the environmental impact and help critters find food, their way home, or even a mate. You wouldn't want to stand in the way of true love would you?!

It's not just clever solutions like these that are needed to protect ecosystems. As consumers, *we* have some influence on the market. Purchasing toilet paper and wood products from traceable and sustainable sources is one easy way we can ensure habitat protection. If you have a garden or access to some kind of outdoor space, planting seeds that are wildlife friendly can provide support for insect populations – and never use artificial fertilisers, or those that contain peat. Vote for people who genuinely care about the land they're representing. This is not just important in the countryside either. Cities can be adapted by planting more trees and by not having so much cement and tarmac everywhere. Sand and soil are better, as they absorb water and keep cities cooler in the increasingly inevitable (thanks to climate change) heatwaves. Again, good for humans AND good for the environment.

We can also all play a small role in keeping an eye on biodiversity in our local area. Citizen scientists have interest in all kinds of subjects from astronomy and biology to environmental and social sciences. Connect with like-minded individuals in your area, or see if there are clean-up or allotment projects you can get involved in. Non-scientists can help in biodiversity tracking, where volunteers can play a major role in monitoring biodiversity by counting and observing species of plants, insects, birds and animals. There are loads of projects going on, and they vary from area to area, but here are three big ones to get you started:

- Sightings of jellyfish and other marine organisms:
  https://www.jellywatch.org/
- The Big Butterfly Count UK:
  https://bigbutterflycount.butterfly-conservation.org/
- BirdWatch UK: *https://www.bto.org/our-science/projects/gbw*

In the immortal words of The Bloodhound Gang: "You and me baby ain't nothin' but mammals, so let's do it like they do on the Discovery Channel…" by protecting our home planet and fellow species, of course! Why? What were you thinking of?

CHAPTER FOUR

AIR

THE GREENHOUSE EFFECT

THE GHG GANG

Here's a fun fact about the greenhouse effect: it is necessary for life. Without any of the natural consequences of the greenhouse effect, the Earth's average temperature would actually be around -18°C because of its distance from the Sun. That's as cold as your ex's reply to your 3am drunk text.

It's thanks to the cosy insulating properties of water vapour, carbon dioxide and the other greenhouse gases that the Earth is habitable for humankind or our non-human kin. However, the line between insulation and a runaway greenhouse effect is thinner than a 1990s pencil brow.

To get technical, gases such as carbon dioxide, methane, water vapour, nitrous oxide and ozone "trap" (by absorbing) the sun's warming infrared rays in the lower atmosphere. Collectively, these are the infamous greenhouse gases (or GHGs on the streets). So the sun rays – or thermal infrared radiation if you want to be specific – keep the Earth warm. As part of the natural process, the Earth reflects some of this infrared radiation back off its surface. However, the GHG gang traps a percentage of this radiation in our atmosphere, absorbing it in their cheeky gases to retain the heat on Earth.

WHAT A LOAD OF BULL

You've probably heard of methane. It might even be the reason you went vegan for a week that one time before caving in to bacon as soon you smelled that fry-up. Well, when it comes to the GHG Gang, methane

GLOBAL GREENHOUSE GAS
EMISSIONS BY SECTOR (2016)

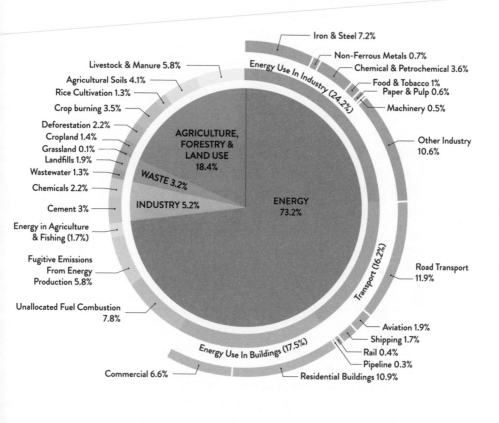

Iron & Steel 7.2%

Non-Ferrous Metals 0.7%

Chemical & Petrochemical 3.6%

Energy Use In Industry (24.2%)

Food & Tobacco 1%
Paper & Pulp 0.6%

Machinery 0.5%

Livestock & Manure 5.8%

Agricultural Soils 4.1%
Rice Cultivation 1.3%

Crop burning 3.5%

Deforestation 2.2%
Cropland 1.4%
Grassland 0.1%
Landfills 1.9%
Wastewater 1.3%

Chemicals 2.2%

Cement 3%

Energy in Agriculture
& Fishing (1.7%)

Fugitive Emissions
From Energy
Production 5.8%

Unallocated Fuel Combustion
7.8%

Commercial 6.6%

AGRICULTURE,
FORESTRY &
LAND USE
18.4%

WASTE 3.2%

INDUSTRY 5.2%

ENERGY
73.2%

Other Industry
10.6%

Transport (16.2%)

Road Transport
11.9%

Aviation 1.9%
Shipping 1.7%
Rail 0.4%
Pipeline 0.3%

Energy Use In Buildings (17.5%)

Residential Buildings 10.9%

is one of the worst. Also known to scientists as CH4, methane forms naturally in the process of decomposition in wetlands, in organism digestion (farts and burps, to the layperson), and in the natural processes of oil and gas formation. While there is far less of it in the atmosphere than CO_2 – something we are all thankful for – CH4 is dangerously powerful, retaining far more heat per tonne than the OG, carbon dioxide. A whopping 25% of global warming is caused by methane, with much of it due to humans being pretty selfish with our steaky desires.

However, the beef with CH4 doesn't stop at cattle farming. Here's a breakdown of the worst CH4 culprits:

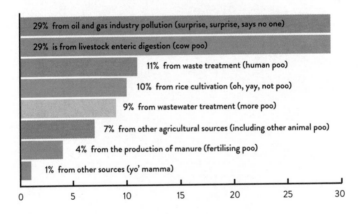

29% from oil and gas industry pollution (surprise, surprise, says no one)

29% is from livestock enteric digestion (cow poo)

11% from waste treatment (human poo)

10% from rice cultivation (oh, yay, not poo)

9% from wastewater treatment (more poo)

7% from other agricultural sources (including other animal poo)

4% from the production of manure (fertilising poo)

1% from other sources (yo' mamma)

0   5   10   15   20   25   30

Here's the kicker: Remember when we said methane was powerful? Well, CH4 has a global warming potential **84 times greater** than CO_2. Specifically, that means that in a twenty-year time frame it will retain 84% more of the Sun's rays than CO_2 would. Worryingly, because the Earth's temperature is already rising, the tundra –permafrost that covers roughly one-fifth of the Earth's land surface – is thawing. Why is this important? Because the vast tundra contains even vaster quantities of CH_4. So, because of human-made emissions of CH_4 (and other GHGs) warming the planet, the tundra is melting fast, letting its own trapped CH_4 join the party. As CH_4 causes more damage than CO_2 over a shorter time frame, this is apocalyptic levels of worrying.

WELL, WELL, WELL, IF IT ISN'T CARBON DIOXIDE...

Of all the GHG gang, carbon dioxide is the best known. Much like vodka and mixer, when measured correctly, CO_2 is delightful. Essential, even. But too much of a good thing and the vodka lands you on the floor asking an umbrella if it's angry at you. Basic natural levels of CO_2 are good, but as with the vodka, too much is disastrous. Just like with the tundra-melt it can cause a disastrous feedback loop: the more CO_2 that gets trapped in the atmosphere, the more heat it traps, which in turn cause more CO_2 emissions. It's like a gassy version of that awful multi-level marketing rubbish ("it's not a pyramid scheme!") your gullible cousin keeps pushing.

However, variations in carbon dioxide are not new to this planet. As of January 2020, CO_2 had a global average concentration of 412 parts per million by volume (or 622 parts per million by mass), which is 50% more than it was before the Industrial Revolution. But from carbon dating in fossils, we can measure that 500 million years ago (around the time the goddess Cher first came to fame) the CO_2 concentration was in fact 20 times greater than today. Even during the Jurassic Period it was five times what it is now. However, normally changes on this scale occur over literally millions of years, allowing life on Earth to adapt. Not hundreds of years! The fact that humans can mess up carbon dioxide this much so fast is almost impressive. Its not, though, it's really terrifying. Like, c'mon, we've all seen *Jurassic Park* – don't go back to dinosaur times!

THE GOOD NEWS: THE OZONE LAYER

Much like shell suits, the ozone layer was a big deal in the '90s.
So cast your mind back to the days of the Spice Girls and Tamagotchis,
and follow us on a heart-warming tale about something
humanity actually got right.

The ozone layer is a layer of atmosphere that protects humans from the
Sun's harmful high-energy UV radiation. Too long an exposure to UV rays
is super-dangerous. Not only do the rays cause skin cancer and immune
deficiencies in humans, but they also affect the balance of ecosystems by
reducing levels of plant growth. The ozone layer is our friend. It protects
us by bouncing these harmful UV rays back into space.

By the 1960s and 1970s, western industrialised nations started using
chemicals called CFCs (chlorofluorocarbons) in products such as
refrigerators and plastics. It was thought that CFCs were a great idea at
the time; they were cheap, non-toxic and didn't catch fire easily. However,
they messed with our buddy, the ozone layer. The chemicals released into
the atmosphere started burning a hole in the ozone layer right above the
Arctic (and other places), letting the UV rays in for a free ride.

But we actually *did* something about it!! In 1987, the Montreal Protocol
was enacted to ban the use of CFCs. Eventually, our friend the ozone
layer began to heal and close back up. Good for you, pal.

The ozone layer today is a fabulous example of how international
cooperation and actual actions rather than just lovely words can ensure
positive change and recovery in the environment.

"THE HOLE IN THE OZONE LAYER IS A KIND OF SKYWRITING. AT FIRST IT SEEMED TO SPELL OUT OUR CONTINUING COMPLACENCY BEFORE A WITCH'S BREW OF DEADLY PERILS. BUT PERHAPS IT REALLY TELLS OF A NEWFOUND TALENT TO WORK TOGETHER TO PROTECT THE GLOBAL ENVIRONMENT."

Carl Sagan, American astronomer, planetary scientist, cosmologist, astrophysicist, astrobiologist, author, poet and all-round smart person.

WHAT'S THE SOLUTION?

It's renewable energy, baby! So fresh, so clean, so goddamn sexy. Where does all this gorgeous renewable energy come from? The whole point of renewability is that energy is taken from sources that can be naturally replenished, such as solar, wind, tidal, geothermal, waves, etc. The good news is that renewable energy is already widely available and being used for electricity generation across transportation, off-grid energy and temperature regulation. The bad news is that we're not doing enough.

Having a world that runs fully on renewable energy isn't just a pipe dream anymore. In fact, Iceland and Norway are already doing it. These nations generate nearly 100% of their electricity using renewable energy. Norway actually funded this transition using oil funds (you love to see it helping for once). Iceland uses mainly geothermal activity, using the country's volatile volcanic situation to their advantage. Every country will have its own unique challenges, but they have proved that with a bit of investment and planning, it is eminently do-able.

LET'S LOOK AT THE FACTS

- Globally, the renewable energy sector is estimated to provide 7.7 million jobs worldwide. Solar energy (or photovoltaics) is the largest employer.
- Renewable energy systems are becoming increasingly efficient and cheaper, and the global share of total energy consumption that is obtained from renewable sources is increasing.
- Between 2019 and 2021, over two-thirds of global newly-installed electricity supply was renewable.
- As of 2019, thirty nations have renewable energy contributing over 20% of their supply, and national renewable energy markets are growing along with demand.

• Renewable energy resources exist over wide geographical areas, meaning that access to energy can be geographically more local. Fossil fuel sources, on the other hand, are concentrated in a small number of countries who monopolise the market.

WIND AND COAL

Wind farms are a fascinating example of how to effectively mitigate current emissions. Not only do they do the same work that burning coal and gas does (providing sweet, sweet energy), but they also result in less greenhouse gas being released, making the effects of burning the coal that we currently *have* to use – for example to make steel – far less severe.

However, it can be hard to see the turbines from the trees, especially in one of the many towns with a history connected to coal mines or plants. In places such as the north of England and parts of Wales – and all over the world – communities are profoundly emotionally invested in industries such as coal mining. For decades, coal plants have both powered huge percentages of people's power needs and driven the economy in these areas. The fact remains, though, that burning coal for energy is far more toxic than other fossil fuels, causing smog, polluting water, causing asthma and cancer – on top of the standard greenhouse gas emissions.

It is important that when moving towards a world where we no longer depend on fossil fuels, we provide economic boosts to areas left hurting by the collapse of the jobs in these industries. The most obvious way to do this is to create jobs in renewable energy industries – such as wind farms – in the same geographical markets as those jobs that are leaving. That way we can not only literally save lives taken by toxic pollution, but also jobs and people's well-being.

EPIC FAIL: THE CAPE WIND PROJECT

Renewable energy seems like a no-brainer, right? Surely everyone can get on board with stopping the accelerated choking and heat death of the big space rock we call home! If only. Let's look at a wind turbine project that never came to fruition: The Cape Wind Project.

For 10 years, opposition to this proposed offshore wind farm off Cape Cod in Massachusetts, USA, built up from politicians, local fishermen (who feared the wind farm would damage their livelihoods), wealthy landowners (who complained that the wind farms would ruin their views), and even Indigenous communities (who were concerned about sites of cultural and religious importance). Because of this hostility, in 2017, instead of supplying tonnes of renewable energy and lightening the dependence on dangerous fossil fuels, the Cape Wind Project was abandoned after a decade of work and a whopping $100 million investment. **Epic. Fail.**

Projects such as these –which seem obviously brilliant on paper – are continually being struck down by the NIMBYism (Not-In-My-Back-Yard-ism) of local populations. The lesson is that there is no way that these huge changes can simply be forced on societies who don't want them. It is up to the developers, the scientists, the local politicians and, in fact, all of us, to educate each other on their importance, and accept that some sacrifices might have to be made for the greater good.

It is often at local levels that wind turbines have been met with the staunchest opposition, but we still need to be pushing at a higher level to hold governments to account for allowing these projects to fail. If the past few years have taught us anything, it is that governments don't have all the answers (and that pyjamas are perfectly acceptable office wear, as is cereal for dinner. It's called adulting). Agencies like the International Energy Agency (IEA) are an international bodies working with countries to discover those solutions and to shape sustainable energy policies.

KEEP GOING FORWARD...

In 2018, the IEA released a report showing what proper investment in renewable energy could do. We're not going to lie, it's delicious. By 2040, the report suggests that the share of renewables could almost double, from 26% today to 44%, surpassing coal as the main use of energy as early as 2026. Solar and wind energy could rise from 7% to 24% of global energy generation, causing dependency on fossil fuels to fall.

However, none of this will happen if projects like The Cape Wind Project are allowed to fail. The technology to achieve these changes is there, but how can we help on an individual level? Find your voice: go on marches in support of renewable energy, vote for politicians who support green policies and, importantly, think local. It's likely that there are proposals

"THE CONCEPT OF GLOBAL WARMING WAS CREATED BY AND FOR THE CHINESE IN ORDER TO MAKE U.S. MANUFACTURING NON-COMPETITIVE."

Ex-President Donald Trump showing off his limited scientific – and geopolitical and economic – understanding.

like the Cape Wind Project around your area. Find out what they are and
how you can lend your weight in support. Even if it is just walking around
your area to inform people of the project, you will have done your bit.

Using renewables does not have to mean living in caves and making your
own clothes – as nice as that sounds for a weekend at Glastonbury, nobody
wants to live their life like that. It's important that people understand that
we will never be able to "go back" to life in the past. We need to change our
expectations, pace of life and priorities to generate solutions to create new
ways of living. Despite what some climate sceptics say, renewable energy
will not result in blackouts but rather cleaner air, water and better political
policies. These are all positive changes, people just need to hear about them!

WORLD LEADERS IN RENEWABLE ENERGY (2018)

Hydro, wind, solar and geothermal as a percentage of total electricity production

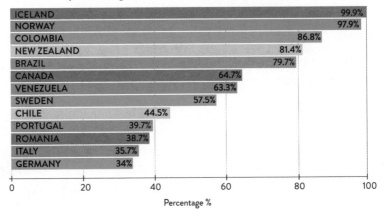

Country	Percentage %
ICELAND	99.9%
NORWAY	97.9%
COLOMBIA	86.8%
NEW ZEALAND	81.4%
BRAZIL	79.7%
CANADA	64.7%
VENEZUELA	63.3%
SWEDEN	57.5%
CHILE	44.5%
PORTUGAL	39.7%
ROMANIA	38.7%
ITALY	35.7%
GERMANY	34%

OILY SLIPPERY CUSTOMERS

The oil and gas industry is gas-lighting us.

It is toxic on so many levels. It's like someone's deadbeat ex telling them: "Babe, you still need me!" But in this case it's to the entire world. Here's the thing – we actually don't need them anymore (or that awful ex – stay strong!). Sure the break-up will be messy and the world *will* spend weeks eating ice cream in front of *Notting Hill*, but in the end it will emerge a strong and beautiful butterfly who is so much better off without them.

If any single sector can change the fate of our planet, it's the oil and gas industry. They have the power to change the course of almost everything. Instead, they spend millions of dollars on ads telling us how much we depend on petrochemical products such as plastics, chemicals, fuel and pharmaceuticals. They adore boasting about how essential they are to all social, industrial and economic systems blah blah blah....

Here's the thing, while propping up those systems that they deem "essential", they are polluting and devastating the very planet we all live on, and all the while they are spending more money than the average country's GDP on lobbyists to pressure politicians into not regulating them, think tanks to produce reports letting them off the hook and attacking alternative forms of energy, and scientists willing to sell-out and say they aren't so bad actually, once you get to know them. It's short-sighted, it's wrong, it wastes our time when we could be developing alternatives to replace them, and it's time we had a talk, petrochemical industry.

Here's one fact that sums up their approach. In 2005, BP spent $250 million on an ad campaign to popularise the term carbon footprint. Was this some altruistic move to try to heal the world? What do you think? **No!** They wanted to shift the blame from themselves – the company who pump untold amounts of greenhouse gases into the air ever year and would literally spill 5 million barrels of oil into the ocean in 2010 (see page 108) – onto you – the consumer just trying to live your own ethical life. Is it any wonder that you feel guilty when you buy an apple imported from Spain rather than one that is home-grown? It was utterly despicable and unfortunately it worked.

Now it is time for them to face a reckoning – and to take some fricking responsibility.

DIRTY MONEY

The oil and gas industry is not just rich, it's **SUPER** rich. Marvel comic super-villain levels of rich. It's also, shockingly (lol not), the largest polluting sector in the world in terms of emissions and most other forms of pollution.

If it was a good global citizen, perhaps this could be partially forgiven, although perhaps not. But except for in a couple of state-owned instances (such as in Norway) the billions that it rakes in never get reinvested in essential public services such as quality education for all, public transport and infrastructure, public housing or healthcare for all. We need to question who this revenue is for? It certainly ain't us anyways.

To make matters worse, despite being worth billions, governments still, to this day, **hand out subsidies** to the oil and gas industry. That's right. Despite being toxic, worth billions of dollars and completely unrepentant, governments still help to pay for part of production costs by offering tax credits or reimbursements. **WHY?** Could it have anything to do with the army of lobbyists and campaign funds they deploy in centres of power like Washington D.C. and Westminster? These industries are literally killing us, and we are paying them to do it!

Angry yet? Be prepared to get angrier. Here's two quick examples where the oil industry has interfered in policy, decency and – we hate to pass judgement here – general morality.

HFO IN THE ARCTIC

Heavy fuel oil (HFO) has a tar-like consistency that causes more pollution and emissions than other oils. When burned, HFO emits black carbon, a short-lived pollutant that is the main component of soot, absorbs sunlight and traps heat in the atmosphere, and is among the most carcinogenic and lethal to humans of all air pollutants. It's not really the sort of thing you like to have hanging around.

The Arctic, which is already warming twice as fast as the rest of the world, is particularly sensitive to these emissions because it darkens the snow and ice, reducing albedo – its ability to reflect solar radiation back into space.

At the moment, HFOs are used in shipping because it is cheaper than cleaner counterparts. In 2011, the UN took the reasonable step of banning HFO use in the Antarctic to protect its sensitive environment. While laudable, this was a very easy step to take because there are no regular cargo routes through the Antarctic. So thanks UN, it's a lovely gesture. Now surely they will do the same for the Arctic, where there is actually shipping, right? *Right?*

To be fair, they are currently trying to pass similar legislation for the Arctic, which if passed would come into effect in July 2024, only 13 years late. At least it would... if only oily Russian oligarchs (oops, I mean politicians) and the petrochemical industry stopped interfering.

Thanks to pressures from the Russian government and oil industry, a host of exemptions and waivers are being put in place that would allow roughly three-quarters of ships using and carrying HFOs to continue to pollute Arctic waters until 2029. According to a study by the International Council on Clean Transportation (ICCT), as the Arctic fleet grows, so will the number of ships that qualify for an exemption, "and the effectiveness of the ban would be further eroded."

And the numbers of ships *will* grow, because melting sea ice (umm... it's melting for a reason everyone... hello?) actually opens up more shipping routes in the Arctic. In this absence of regulation, HFO use in the Arctic is rapidly increasing. Between 2015 and 2019, its use by oil tankers rocketed by 300%, according to the ICCT. While Russia claims that the ships are essential for supplying Arctic communities, researchers and activists doubt that this is the case.

The time for change is running out. Researchers found that if the draft of the ban as it is now with the exemption carved out by the Russians had been applied in 2019, it would have reduced the carriage of HFO by only 30%, its use by 16% and emissions by 5%. Doing away with the exemptions and limiting waivers would have reduced HFO use by 75% and cut black carbon emissions by more than a fifth. For now though, thanks to the sticky hold oil has over governments, it's full steam ahead for black carbon and the steam coming off of all that melting Arctic ice for at least another decade.

CANADA'S TAR SANDS

There's more to Canada than moose, maple syrup and hockey. Did you know that the Great Wild North is one of the world's largest oil producers? And it's all down to the tar sands of Alberta. It's the most destructive industrial project in the world, if Alberta were a country it would be the fifth-largest producer of oil.

Of all the ways to make oil, tar sands are one of the WORST for the environment. Firstly, ancient boreal forest is razed to the ground in order to extract bitumen (a type of hydrocarbon) from *massive* pits. Seriously, it's massive. Just look at Fort McMurray on Google Earth to see how enormous it is. Secondly, the process of turning bitumen into oil is hugely energy-intensive and destructive, causing devastating air and water pollution.

Despite appearing to promote action on climate change on the world stage, the Canadian government continued to not only support, but actively encourage this industry of mass destruction. In 2018, they forked out a whopping US $3.4 billion to buy the Trans Mountain Pipeline, a pipeline that carries oil to Canada's west coast from the Alberta oil sands, because Texas-based owner Kinder Morgan were trying to nearly double its capacity to 890,000 barrels per day, and were having trouble with protests from Indigenous communities and environmental activists, who were taking them to court to stop them. This purchase was not made to help stop the pipeline. In fact it was the opposite: a taxpayer-funded bailout of the oil industry to ensure work continued in the face of this stiff opposition. Construction is still continuing in 2021.

Meanwhile, Indigenous communities – already suffering from the legacy of colonialism – are continuing to experience grave problems because of the tar sands. Medical professionals have been increasingly worried about the rates of cancer, stillbirths, miscarriages and other serious health problems in their communities caused by pollution from the tar sands. Their complaints were dismissed by the Canadian Supreme Court in 2020.

Not only will the new pipeline cross Indigenous lands but it will also run through the world-famous Jasper National Park. When the end point is reached on Canada's western coast, it will dramatically increase the number of oil tankers pulling up along the fragile coast, risking harm to the already endangered orca population and increasing the chance of destructive oil spills.

From the destruction of the boreal forests and their ancient ecosystem, to the damaging high-intensity thermal process to transform it to oil, to the long-lasting toxic chemical contamination of into the land and water surrounding it, to the poisonous effects on the largely Indigenous communities around it, to the harmful impacts of the transportation of the oil, and finally – lest we not forget – to the emissions caused by the actual use of that oil, Canada's tar sands are a noxious example of how companies get rich from damaging activity, and the minute the going gets tough, governments are happy to swoop in and use taxpayer money to bail out those companies and keep the despoiling industry going. Phew, that's a long sentence, but it's only because it's so enraging!

WORST ACCIDENTAL OIL SPILLS INTO THE OCEAN

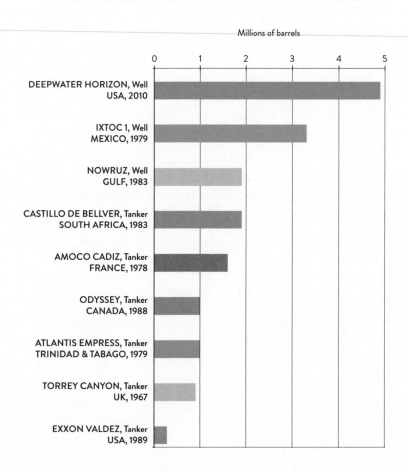

Millions of barrels

| | 0 | 1 | 2 | 3 | 4 | 5 |

DEEPWATER HORIZON, Well USA, 2010

IXTOC 1, Well MEXICO, 1979

NOWRUZ, Well GULF, 1983

CASTILLO DE BELLVER, Tanker SOUTH AFRICA, 1983

AMOCO CADIZ, Tanker FRANCE, 1978

ODYSSEY, Tanker CANADA, 1988

ATLANTIS EMPRESS, Tanker TRINIDAD & TABAGO, 1979

TORREY CANYON, Tanker UK, 1967

EXXON VALDEZ, Tanker USA, 1989

DEEPWATER HORIZON SPILL

You know the old saying, there's no point crying over spilt milk? Well that doesn't apply to oil. There is very much a need to cry over spilt oil.

Oil spills damage ecosystems and practically all life on earth in the short, medium and long term. It's a triple threat! Not only does it cause mass death of both land and sea animals, but once the oil sinks from the surface of the water to the ocean floor it continues to poison ecosystems for up to 100 years.

In the Gulf of Mexico on 20 April 2010, there was an explosion on the BP-operated Macondo Prospect. The disaster was when methane burst through a recently installed cement well cap meant to seal a well drilled by the Deepwater Horizon oil platform. When the gas burst through, it ignited, instantly killing eleven workers and injuring seventeen. The platform capsized and sank slowly over then next two days.

It took several months for the well to be capped, but by then the damage had been done. Between 130 and 210 million gallons of oil was spilled, coating an area of over 2100 km^2 (1300 mile^2) in oil. Only about 25% of the oil was actually recovered – the rest is still out there, doing long-term damage.

This was a disaster for wildlife in and around the Gulf of Mexico. The oil spill harmed or killed millions of animals, from birds to marine mammals such as bottlenose dolphins, spinner dolphins, melon-headed whales and sperm whales. Just using one species as an example, the US government's National Oceanic and Atmospheric Administration estimate that approximately 4,900–7,600 adult sea turtles and 56,000–

166,000 small juvenile sea turtles were killed by the spill. Depressingly, even some efforts to clean up the oil spill made it worse. Nearly 2 million gallons of toxic dispersants were sprayed into the Gulf's waters, worsening the situation by causing small oil droplets to sink to the bottom of the ocean, damaging deep-sea ecosystems.

BP paid $65 billion in compensation to people who relied on the Gulf for their livelihoods, but nothing can undo the permanent and irreversible damage to the environment, or compensate for the cost of the land and marine life it destroyed. It does not appear as if this fine was big enough to teach the industry a lesson, either. In the decade since Deepwater, there have been more than twenty spills of 1,000 tonnes of oil or more. Until we replace oil with a cleaner fuel, these deadly incidences will keep occurring.

"I THINK THE ENVIRONMENTAL IMPACT OF THIS DISASTER IS LIKELY TO HAVE BEEN VERY, VERY MODEST."

Tony Hayward, **the CEO of BP**,
minimising the carbon footprint
(*remember that one, Tony?*)
of the Deepwater Horizon disaster.

DENIED
DENIER

PHASING OUT OIL

In 2020, a newspaper headline from *The Guardian* asked "Will the coronavirus kill the oil industry?" They weren't the only ones wondering. The COVID-19 pandemic changed a lot of things: toilet roll was gold dust, banana bread became an actual currency and oil prices reached a historic low. Global lockdowns and travel bans saw airplanes grounded and in many areas there was a marked improvement in air quality. The pandemic showed that it was possible for the environment to recover, that oil isn't the behemoth it likes to think it is and that the maximum number of times one should wear the same sweater is exactly 17 days. However, within a few months, emissions were already on the rise again, and were expected to regain their pre-pandemic peak by the end of 2021. So how can we make the short-term reduction a long-term reality?

Phasing out oil is a complex business. It really has seeped into most industries in one form or another, and each have their own solutions that would take a whole book to explore just on their own. Very briefly, though, much transport can be electrified (cars, trucks) or powered by hydrogen gas (shipping), which is *not* a member of the GHG squad. While plastics are currently everywhere, we can reduce how much we use single-use plastics and find alternate materials for products and packaging that currently need it.

What the myriad of solutions all need, though, is the investment that has so far been lacking. All of that might soon change, though. Economically speaking, oil is not seen as the safe investment it once was. The industry has lost its appeal – it's been over-supplying global markets for a number of years, driving the price down, and with more and more countries pledging to become carbon neutral by 2050, investors can see the writing on the wall.

In 2019, the oil industry's share of the S&P 500 stock index hit its lowest level in four decades. On CNBC, *Mad Money's* Jim Cramer declared that fossil fuel stocks were in their "death knell" phase, and that was before anyone had even heard of a lockdown! It is becoming clear that the oil and gas industry has failed to reconcile its business model with climate reality, and instead it is doing its best to create new markets for its products and suppress competition from renewables while planning for a future filled with oil and gas. In fact, the industry has already invested in pumping out more oil and gas than the world can afford to burn and stay under its climate targets, yet it keeps investing in pumping out even more. It is burying its head in the (tar) sands.

Making oil unviable from an economic point of view is the single biggest thing that can be done to turn the tide. Some of this is happening naturally as alternative energy prices drop below oil and gas prices, forcing oil prices ever lower, to a point where they are economically unsustainable. But we also need governments to stop subsidising them, and even worse, bailing them out (Trudeau, you really let us down in Canada... and you seem like such a nice guy. What happened?). When this happens, they will have no choice but to turn their immense wealth to green energy simply to stay in the game, and then they can be part of the solution instead (and make sure they can be leaders in the green energy market too, cheeky). Some companies like Equinor in Norway are already doing this, and this is the sort of greenwashing that we should encourage!

As the world recovers from COVID-19, governments must make the decline of oil and gas a priority. Returning to pre-COVID "normality" is a terrible idea for the energy industry. Phasing out oil should be part of any "new normal" plan. The oil industry is dying a natural death. Let's unplug that life support machine.

TRANSPORT:
THE SOLUTIONS

Phasing out oil and gas is an excellent plan, but ultimately we need new solutions for the areas that depend on it the most, and transport is the most obvious. Transport systems emit enormous volumes of greenhouse gas. There are currently an estimated 1.4 billion cars and 25,900 commercial airplanes in use, not to mention motorbikes, buses, trucks, ships... you get it, a lot of stuff we need every day depends on oil.

THANKS, IT'S ELECTRIC

Electric cars are the talk of the town when it comes to renewable transport solutions. On paper, they seem like a fantastic idea, just charge 'em up and go. What's not to love?

Electric vehicles (EVs) consume about a third of the amount of the energy used by vehicles with an internal combustion engine, making them cleaner and more efficient than cars that are powered by fossil fuels. EVs are still more expensive than petrol, but between 2010 and 2018 the cost of EVs fell by 84%. According to the Electrical Energy Agency, by 2018 5.1 million EVs were being used on the roads compared to just 2 million in 2017. In 2019, the majority of car sales in Norway (these guys again... seems like they are doing something right) were of EVs. This is a **GREAT** development. Be they hybrid or fully electric, EV's are unequivocally better for the environment than conventional cars.

However, how beneficial would EVs be if you had to charge them up via a coal-powered grid? Not at all, which is why renewable energy is just as important to clean up this sector as replacing petrol cars.

GREEN TRANSPORT PYRAMID

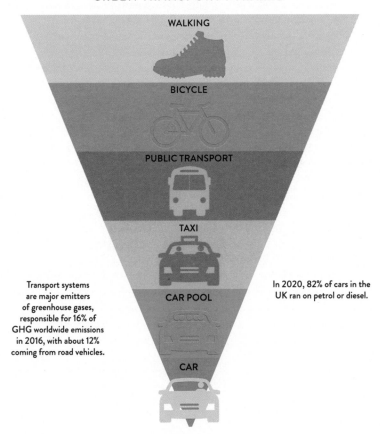

WALKING

BICYCLE

PUBLIC TRANSPORT

TAXI

CAR POOL

CAR

Transport systems are major emitters of greenhouse gases, responsible for 16% of GHG worldwide emissions in 2016, with about 12% coming from road vehicles.

In 2020, 82% of cars in the UK ran on petrol or diesel.

A trickier argument is employed by some deniers. They will try to tell you that the rare earth materials needed for the batteries actually make EVs *worse* for the environment. These materials need to be mined, and mining both damages the climate and exploits the people who have to do it, especially because they are only found in particular parts of the world where climate and human rights abuses are endemic in the mining methods used. (Sidenote: funny how deniers suddenly get so concerned when the mining hurts the fossil fuel industry, isn't it? But the two centuries of mining since the Industrial Revolution have been just *terrific*.)

For example, cobalt is a crucial material and is very dangerous for anyone to handle, 60% of it comes from the Democratic Republic of Congo, where cobalt mines reportedly use child labour. This is obviously horrific and has to be stopped ASAP – lawsuits have already been brought with that purpose.

These are genuine public health and climate issues and illustrate the complexity of the problem: just because we've solved one issue doesn't mean we've fixed the whole system. However, unlike the petrochemical industry and other climate change deniers, green activists and companies are not burying their heads in the tar sands. Instead, investments are being made to both better recycle and reduce the use of materials like cobalt, and to find new ways to make efficient batteries without them, and the industry is actually welcoming regulations from governments to help drive the process, rather than hiring crews of lobbyists to convince governments to let them off the hook.

OTHER TRANSPORT

We know not everyone can afford an EV, but there are already so many ways to curb emissions. The use of public transport is super-important, so do your bit for the Earth and take the bloody bus! A five-minute wait at a bus stop is a small price to pay for Mother Earth. In many cities, buses themselves are going electric, too. Cities like London, Tokyo and Paris have also introduced low emission zones to help curb the amount of vehicles in city centres and improve the health of the city's people.

Depending on where you live, public transport isn't an option for everyone. But walking or cycling (or jogging, if that's what floats your boat), even car-pooling, for shorter journeys is healthier both for you and for the environment.

Shifting our focus away from buses and cars, the transition to zero emissions gets much trickier. Currently, electric engines cannot support trucks, ships and planes because the batteries do not offer the range to cover very long distances. Energy density is an issue for batteries, and if these modes of transport are to be electrified batteries need to be able to store more energy in smaller and lighter batteries, and to charge more rapidly.

Until that tech comes – and we are sure that it will – some sustainable biofuels could help in the short-term, but these still result in emissions. Liquid hydrogen may well be the answer. It comes with many of its own complications, but it has a higher density and fewer potential risks than natural gas and, crucially, pretty much zero emissions! The idea for using it in vehicles has been around for decades, and it is already the fuel of choice in space.

If it can get humans to the Moon, surely it can power us over to Ibiza…

CHAPTER FIVE

LAND

INDUSTRIAL FARMING

Food, glorious food, where the hell does it come from? At the most basic, primitive level, humans need food – "well, *obviously*", we hear you say. Eating is such an intrinsic part of the human experience, whether it's an elaborate three-course meal or cold pizza for breakfast (we don't judge). From sushi to sandwiches, broccoli to beer, everything we use to fuel our bodies has consequences – not just for us and our digestive health, but for our planet.

Every bit of food produced is intricately connected to nature. Much like fishing (for more on oceans, see Chapter 6), modern farming techniques are bringing ecosystems and the climate to crisis point. How we currently feed the world is degrading soil quality, ecosystems, depleting water resources and worsening an already dire climate situation. According to the United Nations Food and Agriculture Organisation (the FAO), food systems (including the production of food and non-food items like leather, land use, agriculture, refrigeration and packaging) accounts for **over one-third of global greenhouse gas emissions**.

Old MacDonald sold out and we are not impressed.

We're currently pushing the limits of what we can produce. The main driver for modern food production is profit: cheapness at whatever cost in order to maximise profit. The production of more food, more cheaply, has resulted in chemicals, pesticides and animal by-products leeching into ecosystems. Current practices indicate that food production is expected to increase by 70% by 2050, further stressing soil quality. Yet the available land suitable for agriculture remains unchanged.

CLONE WARS

Star Wars: Clone Wars, The Stepford Wives, Never Let Me Go, there have been sooo many books, TV shows and films about why lots of clones are a VERY BAD IDEA.

Yet, many industrial farms use monocultures (genetically uniform crops) as a means to try and save money. This is super risky as monocultures (such as the Cavendish banana – see below) are very vulnerable to pathogens and pests due to a lack of lack of genetic diversity. In order to keep these fragile crops healthy, farmers then waste money and resources on more antifungals and insecticides, which can be devastating for ecosystems. Crop/livestock viruses (such as Panama disease in bananas) evolve quickly and easily against monocultures and can decimate entire crops or herds. As we all know from bird flu, swine flu, HIV and COVID-19, many of these viruses then make the jump into human populations.

One traditional response has been to switch one monoculture crop out for another that is more disease-resistant. However, specialists have suggested this is merely a short-term solution because pathogens evolve faster than farmers and GM (genetic modification) technology can keep up.

Let's look at the bananas mentioned above. There are many varieties of banana but 99% of consumed bananas are a variety called the Cavendish. However, the main banana crop of choice used to be a variety called the Gros Michel – which is now extinct! In the 1950s, the entire world's crop was ravaged by a strain of Panama disease which quickly spread through the clone monoculture population. Farmers had no choice but to switch to the Cavendish, which was fortunately resistant to the disease. There is now no back-up and, as a result, the contemporary banana supply chain is at risk of collapsing – in fact, it's not a matter of *if*, but *when*.

Mixing crop strains to avoid monoculture farming is not only the best way to avoid a mass crop extinction, but it is also the best way of removing pests from a main crop, while reducing pesticide reliance and protecting soil health. As well as this, different crops have different root depths allowing a more varied and healthy use of soil – helping worms and insects do their thing.

THE BEE'S KNEES

We need to save the bees! Not just because they've got cute little fluffy butts (Google it right now) but life on earth would literally collapse without them. Bees pollinate a third of ALL the food we eat, including vegetables, coffee and fruit. These busy little angels also pollinate foods (berries, fruits and nuts) that wild mammals and birds eat, making them essential to food webs. Honeybees alone pollinate $15 billion worth of US crops every year. They deserve a holiday, pension and sick pay, not to be choked with insecticides.

Bees and butterflies are probably the most studied insects, but the data is depressing. One in ten bee and butterfly species is threatened with extinction in Europe. Over the last few decades beekeepers all over the world have reported bee colony losses. In countries such as the UK, France, Belgium, Portugal, Germany, Italy, Spain and the Netherlands, bee death has been particularly high – but pollinator extinction is a global threat.

In February 2019, a study was published which found that around 40% of honey-bee colonies in the US had perished over that winter. It's not just bees, moths and butterflies are also disappearing; between 2000 and 2009, 52% of UK butterfly species on farmed land disappeared. Unless this trend is reversed, we will find it impossible to feed our current population, let alone the much larger one forecast by the UN in the next few decades, and mass hunger is sure to be the result.

"IF THE BEE DISAPPEARS FROM THE SURFACE OF THE EARTH, MAN WOULD HAVE NO MORE THAN FOUR YEARS LEFT TO LIVE."

This dire warning is popularly attributed to Albert Einstein. You should listen to him, apparently he was a very smart guy.

GLOBAL GREENHOUSE EMISSIONS

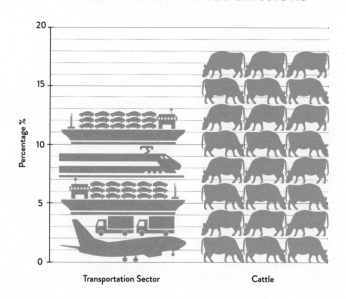

DON'T HAVE A COW

Cattle farming is a load of BS. Literally.

Most food is factory farmed, but in the US this accounts for a whopping 90% of consumed meat, so it is a large component of our consumption. Unfortunately, it comes at a price. Livestock farming alone contributes 18% of our total emissions of GHGs (see graph on previous page). In 2006, the FAO described livestock farming as "...one of the most significant contributors to today's most serious environmental problems". It's time to face the moo-sic.

From pen to plate, every part of the beef industry is a climate disaster. Factory-farmed beef requires twice as much fossil fuel energy input as pasture-reared beef.

Livestock farming is incredibly intensive and ruins the environment on multiple levels. Firstly, livestock require A LOT of land. Between 1980 and 2000 an area over 25 times the size of the UK was converted into new farmland – mostly for cattle – in the Global South; over 10% of this land used to be tropical forest. In the Amazon, current patterns of agricultural expansion for grazing and crops threaten to destroy 40% of existing pristine rainforest by 2050. Not only is this awful for already endangered animals, it reduces vital oxygen production and forest clearances also eradicate essential carbon sinks, releasing gases into the atmosphere.

It's not just grazing land that is required for beef farming; even more land is needed to grow their feed. A 2010 study published by the Royal Society confirmed that animal feed production accounts for around 75% of the total energy consumed in rearing of livestock in factory farming and one-third of the world's crop lands are used just for growing animal feed.

If all this leaves a bad taste in your mouth, just wait for what's in store for your nose. Cattle release methane (remember CH_4?), which is far more potent than CO_2 in trapping sun rays and causing global heating. Livestock farming releases between 37% and 65% of global methane and nitrous oxide emissions. Our atmosphere is being choked by cow farts.

Large-scale cattle farming also causes water pollution. Chemicals (such as nitrogen and phosphorus) used in the farming of their feed, as well as run off from their waste and the excessive pharmaceuticals and antibiotics (which can push bacteria to become more resistant, more infectious, and far deadlier) – ALL OF THIS leaks into natural water courses, contributing to water acidification, killing plants and animals, and even leaving vast "dead zones", which are as bad as they sound.

So what can we do? Well, on an individual level, reducing meat intake is always a good first step. There are many other ways to get protein into your diet that don't, you know, leave the Amazon in ruins. Intensive agriculture has to answer for the climate catastrophe and habitat destruction, so try and support family-run or local farms that use regenerative farming methods where possible.

But not everything is ruined! It is important to mention that it isn't necessarily the cow, but the HOW that is the problem. Small communities can live perfectly sustainably on meat that is not factory farmed. Some organic and restorative livestock farms modelled on ecosystems can help to fertilise soil naturally and are nowhere near as harmful because they see the livestock as **part of an ecosystem**, not just a product. These methods, however, are impossible to scale up and thus expensive. But, if you have the money to pay for meat from regenerative small-scale producers, or a forward-thinking government willing to subsidise them, then go for it! .

FOOD, FARMING AND HEALTH

FOOD INEQUALITY

With all this intensive and large-scale farming, you'd be right in assuming that there's enough food to go around. Well, there kind of is, but alas things are not so simple (or fair). Between unnecessary food wastage and shocking food inequality, there are serious issues around who has the ability to create waste and who doesn't have enough to eat.

According to the UN World Food Program (WFP) we currently produce enough food to feed 11 billion people. This is far more than the current population of the Earth, which, at the end of 2020, was just under eight billion people. However, in 2016, the FAO estimated that around 815 million of the global population (about 10.7%) were suffering from chronic undernourishment.

How the hell is this happening? Globally, 30% of food is currently lost or wasted through a myriad of unfair social and economic systems, exacerbated by climate change.

The FAO's report has been important in informing the international 2030 Agenda for Sustainable Development, which aims to end hunger and all forms of malnutrition by 2030. However, hunger appears to be on the rise. Another report published in 2019 by the WHO, UNICEF, WFP and the International Fund for Agricultural Development (IFAD) estimates that 26.4% of the world's human population is experiencing moderate or severe levels of food insecurity, despite the increasing prevalence of mega farms. This amounts to about 2 billion people experiencing food insecurity.

What are the reasons for this? Conflict is one major cause of malnutrition, with the FAO's report estimating that of the 815 million experiencing hunger, 489 million live in countries affected by conflict.

However, food insecurity also occurs in wealthy nations such as the UK, where a 2016 survey by the Food Standards Agency found that 8% of adults (approximately 3.9 million people) living in England, Wales and Northern Ireland were food insecure, defined as experiencing insufficient and insecure access to food due to lack of funds in the preceding 12 months. The study illustrated that rates of food insecurity were as high as 23% for low income adults and up to 47% for unemployed adults.

Whether in wealthy or impoverished nations, the divide between rich and poor is increasing, and healthy food is not as affordable as highly processed unhealthy foods. Many families cannot afford nutritious foods like fresh fruit, organic vegetables, and healthy, sustainable sources of fat and protein. Foods and drinks high in fat, sugar and salt are cheaper, will keep for longer and are more readily available. It is quite common to find malnutrition and obesity in the same communities and households.

"FAILURES IN FOOD SYSTEMS FUEL HUNGER AND LIMIT LIVELIHOODS."

The UN World Food Program.

Some projections suggest that the world's population could increase to 11 billion by 2100. If everyone is to thrive and be fed without destroying all that we depend on, then our food waste has to change along with our supply chains and the everyday habits of people living in industrialised wealthy nations.

FOOD WASTE

Food standards come in two forms: those that make something safe to consume – which makes sense – and random food beauty pageants. This ain't *Toddlers and Tiaras*, you don't need a perfect looking carrot or apple.

The global food system is highly industrialised. Supermarkets and their suppliers have really strict and frankly unnecessary uniformity guidelines applied to fruits and vegetables. Only the fruit or veg that meet the beauty standards set by the marketing team get packed and sold in most supermarkets. Not every piece of fruit can conform to these ridiculous standards, causing growers to have tonnes of left over, but still beautifully delicious, occasionally "weird looking" fresh fruit and veg they can't sell to supermarkets. Where do these ugly ducklings go? Sadly, it goes straight to landfill – about 30% of all fruit and veg produced globally is wasted before it even gets to the supermarket shelf.

In the UK alone, over three million tonnes of fruit and veg is tossed before even leaving the farm. This wastes all the energy, water and time invested in their growth, and stopping this sort of thing globally could eliminate 11% of all greenhouse gas emission. According to IFAD, over 35 million tonnes of China's total food production (around 6%) is lost or wasted annually. This could feed 30 to 50 million people. Even more shockingly, a 2011 report on food waste by the FAO found that, based on emissions alone, if food waste was a country, it would be the third biggest emitter of GHGs in the world. And this dire situation has NOT got better in the last 10 years.

"THERE'S ENOUGH ON THIS PLANET FOR EVERYONE'S NEEDS BUT NOT FOR EVERYONE'S GREED."

Peaceful protester, lawyer and anti-colonialist Mahatma Gandhi had many wise things to say.

WHAT CAN WE DO ABOUT IT?

FOOD NETWORKS AND ORGANIZATIONS

Consumer choices have a huge role to play in how supply chains are managed. Our choices need to be powerful enough to challenge pointless aesthetic marketing standards –we must advocate for better, less wasteful supply chains with how and where we choose to buy our food.

Lucky for us, there are already clever and dedicated people working to change our perception of food quality and waste. One example of activism through innovation is Imperfect Foods. Based in the US and founded in 2015, Imperfect Foods is an online grocer on a mission to eliminate food waste. They source surplus and rejected foods which are then sold via subscriptions and delivered to your doorstep. Not only is this good for the environment, it's good for your pocket too, as they charge around 20-30% less than usual greengrocer costs. So far, they have saved around 120 million lbs of food that would have otherwise gone to landfill. Similarly, in the UK, Odd Box sources fruit and vegetables directly from farmers that don't meet the supermarket beauty standards and deliver directly to your door on subscription. In France, customers in various supermarkets are encouraged to buy cheaper *"fruits et légumes moches"* by pricing hooked cucumbers and two legged carrots at up to 30% less. Bargain! Examples such as these are occurring all around the world.

Another brilliantly inspirational example of activism and initiative is the Food Recovery Network (FRN). Founded in 2011 by students at the University of Maryland, the FRN is the largest student movement fighting against food waste and hunger. The network began as a way to divert food from the college cafeterias to communities in need. Through the FRN over 3.2 million meals which would have otherwise gone to the landfill have been donated locally to people in need. Through saving food that would otherwise be wasted, 6.8 million pounds of CO_2 was saved and hungry people were fed.

WHAT ELSE CAN WE DO?

While large systemic change is needed, there is still stuff we can do as consumers. While these tips may not apply to all readers, depending on your access to healthy food and purchasing power, if you can begin changing your approach to food and food waste then you've started to make a difference already!

- Plan ahead and consider making meals that can be frozen. The freezer is a treasure! Use it.
- Cut down your food waste; eat what you buy! Most people in the world do not have access to good food, so if you happen to be one of the privileged ones be sure to act responsibly and do not waste food.
- If you run a food business, cut down on your food waste. Consider donating excess food to families and people in your local community who need it.
- Pay attention to where your food is coming from. Is it produced using traceable and environmentally considerate processes? Can you shop organic, regenerative, or local?
- Look out for palm oil in your food and cosmetic products. Palm oil is extremely damaging in the way it's produced and may not be from traceable sources.
- Reduce the amount of meat and dairy products you eat, or if you can afford it purchase your meat and dairy from producers with traceable, transparent and environmentally conscious practices.
- Your dentist is going to love you for this one: cut down on sugar! Sugar uses large amounts of agrochemicals, water and land. Run off from sugar farms in Uganda has been found in the Great Barrier Reef! And did you know that to create one kilo of sugar you need 1,770 litres of water? That's nine gallons per teaspoon! If you can afford it, try replacing the refined sugar you eat with 100% honey. Bees pollinate a sixth of all plants on earth, and honey cannot be farmed using chemicals for fear of harming the bees.

EXTREME WEATHER

Much like our uncle Joe, the weather has become increasingly erratic in the past 20 years. (Seriously, NO ONE wants to see a middle-aged man in Lycra puffing his way up a hill on a carbon fibre bicycle bought with his divorce settlement money.) Natural disasters caused by a more energetic climate have left trillions of dollars of destruction in their wake. But extreme weather is infinitely more serious than Joe's mid-life crisis. Landslides, hurricanes, floods and wildfires have taken a gigantic toll on human and animal populations. Climate change ain't helping. Wait, let's try that again: climate change is making extreme weather a HELL of a lot worse.

LANDSLIDES

Landslides and heavy rainfall have always happened. However, increased construction, soil degradation and the increase in rainfall due to climate change are making them more frequent, massive and even more widespread. Countries in Europe, Asia, South America and Africa have all seen more landslides after unusually intense rainfall.

One of the most devastating modern landslides occurred on 14 August 2017 in Sierra Leona. About 400 people in the mountain town of Regent, on the outskirts of the capital Freetown, were killed by a landslide. Homes were swept away and thousands were made homeless. While a landslide is typically considered a natural disaster, the magnitude of this landslide was linked to anthropogenic factors. The recent uprooting of trees to make way for construction on nearby hillsides made the soil unstable and vulnerable to collapse. In the aftermath, the government of Sierra Leone was accused of not doing enough to tackle illegal construction on the overcrowded hillsides.

As well as this, the intense rainfall that triggered the landslide was almost certainly made much more likely by climate change. While storms and torrential downpours are common in Sierra Leone in the months of August and September, in 2017 Sierra Leone experienced 104cm of rain from 1 July to the day of the landslide. According to the US National Weather Service's Climate Prediction Center this was three times more than usual for the rainy season. Torrential rainfall in the week previous, combined with short-sighted construction practices caused many avoidable deaths and impoverished many more.

A SURVEY OF GLOBAL CLIMATE CHANGE CONCERNS

	Droughts or water shortages	Severe weather, like floods or intense storms	Long periods of unusually hot weather	Rising sea levels
SOUTH AMERICA	59%	21%	12%	5%
AFRICA	59%	18%	16%	3%
USA	50%	16%	11%	17%
ASIA/PACIFIC	41%	34%	13%	6%
MIDDLE EAST	38%	24%	19%	5%
EUROPE	35%	27%	8%	15%
GLOBAL	44%	25%	14%	6%

% Represents respondents who were concerned by climate change outcomes.

RISING SEA LEVELS

Since 1993, global mean sea level rise has hit 3.3mm per year. Between 2010 and 2018, sea level rise increased to about 4.4mm per year. It may only seem like a small amount, but the effects of such a rise have already been devastating. For example, in November 2019, Venice was submerged by an exceptionally high tide – the highest in 50 years.

Coastal nations and communities are particularly vulnerable. As a collection of islands with over 80,000 kilometres of coastlines, Indonesia is one of these at risk countries. The capital, Jakarta (population 10 million!), is one of the world's fastest sinking cities, annually experiencing severe flooding. By 2050, 95% of North Jakarta could be beneath the waves. The land is literally collapsing beneath the city, which combined with sea-level rise could displace many of Jakarta's poorest communities, who tend to live by the edge of the sea wall.

HURRICANES, FLOODING AND STORMS

There are several factors that affect how powerful a hurricane can become. For example, a warmer climate increases storm rainfall, intensity and surge. For every 1°C increase in temperature, the air holds 7% more water vapour, leading to the heavier rains that have been recorded all over the world, increasing the risk of rivers and streams flooding. Heating may even threaten the stability of thermohaline circulation (the circulation of heat and salt density) in the oceans. Ocean temperatures have increased by 1–3°C over the past century and melting land ice has already contributed to sea-level rise, which increases storm surges and coastal flooding. Hurricanes are projected to produce more rain and stronger storms are expected to be more frequent. We are already experiencing this change in hurricanes.

Hurricanes Katrina, Harvey and Maria all broke records with their severity and damaged entire communities who are still feeling their effects today.

HOW HURRICANES FORM IN THE ATLANTIC OCEAN

Sea-level rise from melting glaciers could mean stronger storm surges.

Most major Atlantic storms originate in Africa, as a mix of Sahara Desert heat and humid coastal air creates the tropical atmospheric waves that trigger storms.

EUROPE

SOUTH AMERICA

AFRICA

ATLANTIC OCEAN

Rising sea temperatures could feed the strength of the storms.

WILDFIRES

Wildfires are a natural phenomena, but the intensity of wildfires we're seeing today is not. Those that occur naturally (resulting from lightning or as part of natural cycles) and those that occur due to the negligence of a small minority of individuals are increasingly devastating to the landscape, ecosystems and human communities alike. Parched forest and vegetation, and disappearing water tables have created what is called "tinderbox conditions", allowing usually small fires to grow uncontrollably. In recent years, we have seen the worst wildfires in recorded history.

Wildfires in California and Australia have worsened due to climate change, and as wealthy nations, they have garnered the most international attention (another example of climate inequality – you don't hear nearly as much about the devastating Siberian wildfires). Those areas, and in particular the west coast of the USA, have experienced more intense droughts in recent years, with drier winters and heat waves contributing to drought. Record high temperatures in 2020 led to a more intense wildfire season. In fact, it was the most extreme and destructive wildfire season on record in both locations.

Wildfires cause a vicious cycle. The burning trees and shrubbery cause an increase in CO_2 emissions, which then contributes to the warming conditions that cause more wildfires. An analysis led by Stanford University published by the Institute of Physics found that the "temperature in California during the autumn months had increased by approximately $1°C$". Combine this with around 30% less autumn rainfall over the past 40 years, and you get more extreme autumn wildfires – which have "more than doubled in California since the early 1980s". Even worse, fires are spreading to areas that are not normally affected, leaving communities vulnerable to even more loss of life, homes and livelihoods.

On the other side of the Equator, the Amazon is also burning. The world's biggest rainforest (which spans Brazil, Bolivia, Peru, Ecuador, Colombia, Venezuela, Guyana, Suriname and French Guiana) has experienced a year-on-year surge in fires for the past few years. Fires normally occur around the dry season but have increased due to slash-and-burn methods of clearing the rainforest. Deforested areas are then used to make way for agriculture, livestock, logging and mining. Slash-and-burn depletes the soil's ability to retain moisture, worsening conditions for wildfires. While slash-and-burn activity is generally illegal across the Amazon region, the enforcement of environmental protection has been lax at best, practically non-existent at worst.

In the latest complete year, from 29 August 2019, the Instituto Nacional de Pesquisas Espaciais (INPE) reported over 80,000 fires across all of Brazil, a 77% year-on-year increase. These fires released 228 **megatons** of CO_2 – the highest since 2010. Within a mere 7.7% of the total area burned, an estimated 2.3 million animals died.

"YOU HAVE TO UNDERSTAND THAT THE AMAZON IS BRAZIL'S, NOT YOURS. IF ALL THIS DEVASTATION YOU ACCUSE US OF DOING WAS DONE IN THE PAST THE AMAZON WOULD HAVE STOPPED EXISTING, IT WOULD BE A BIG DESERT."

DENIER
DENIED
DENIER

Brazil's despotic right-wing president, Jair Bolsonaro, has a long history of climate change denial and has taken active steps to remove protections from Indigenous lands and encourage slash-and-burn farming.

WHAT'S THE SOLUTION?

We WISH we could tell you that there was a quick and easy way to fix the weather but, unfortunately, there's only one real hope: cut emissions (which, of course, can only be realised by employing multiple climate solutions and changes in behaviour). We must vote for people who will cut emissions and prioritise climate change (and not do things like open coal mines for short-term gains, as the British government was considering doing in early 2021). We can all employ ways of cutting emissions in our own lives, and talk to people around us to get these conversations to move beyond our online bubbles.

On a more local level, forest maintenance can help. Usually leaving forests to be forests is the way to go, but not with fire season going crazy! During times of intense drought (when the water table is, err, gone and tinderbox conditions are getting more tinderbox-y) the removal of dead trees (AKA fire fuel) and the planting of young native tree species can reduce the intensity and extent of fires. Careful fire management with "controlled fires" or '"prescriptive burning" which burns away the excess tinder that an out-of-control fire would be able to use has also been employed, and is an important future tool.

Forward planning is essential in avoiding unnecessary wildfire deaths. We can stop developing cities into or in fire-prone areas, improve the standard of fire safety in homes and the robustness of power grids. In poorer areas like the Amazon, alternatives to slash-and-burn agriculture need to be developed by investing in, and focusing on, businesses that not only support rural communities but protect Indigenous communities, their culture and knowledge, in order to protect the rainforest.

AUSTRALIAN BUSHFIRES

While wildfires and their resulting devastation tend to capture headlines, none has had the global impact of the 2019-2020 Australian bushfires. Record-breaking temperatures and months of drought linked to climate change facilitated the spread of massive bushfires across Australia's eastern coast.

In 2019, the fire season started early in the provinces of Victoria and New South Wales with drought affecting 95% of the state and persistent dry and warm conditions. Twelve local government areas started the Bush Fire Danger Period two months early on the 1 August 2019, and nine more started on 17 August 2019. The fires lasted from June 2019 to May 2020, peaking between December and February.

Experts called the bushfires the worst wildlife disasters in modern history. This gloomy conclusion is unsurprising, given that **almost three billion animals were killed or displaced**. On Kangaroo Island alone 40,000 were killed. Approximately 18,636,079 hectares (46,050,750 acres) were burned. To put that huge number into context, that's almost as large as the entirety of the European continent.

The human cost was also tragic. It is estimated that 34 people died directly in the fire, and 417 died indirectly as a result of smoke inhalation, carbon monoxide poisoning or suffocation. Over 3,500 homes were destroyed, and the total economic cost is estimated at $4.4 billion.

The real shrimp on the barbie? The Australian prime minister at the time, Scott Morrison, has a history of climate change denial. Morrison caused headlines by brandishing a piece of coal in parliament, telling colleagues to not be "scared" of fossil fuels. As leader of the right-wing Liberal Party and head of the Australian government, he stopped payments to the Green Climate Fund, a UN-backed organisation that assists developing countries hit by climate disaster. When extreme wildfires first broke out in 2019 Morrison played down the effect of climate change, saying that; "They are natural disasters. They wreak this sort of havoc when they affect our country, and they have for a very long time." Accused of not instituting enough policies to mitigate climate change, the government promised $2.2 billion in aid to victims of the wildfires, while Morrison has belatedly and slowly changed his tune on the impact of climate change. But is it too little too late? Only time will tell, but it does seem that Morrison is ill-equipped to push through the changes needed. In the meantime, we hope he gets haunted by the ghosts of all those poor kangaroos.

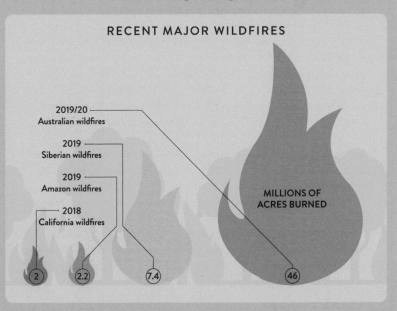

RECENT MAJOR WILDFIRES

2019/20
Australian wildfires

2019
Siberian wildfires

2019
Amazon wildfires

2018
California wildfires

MILLIONS OF
ACRES BURNED

2

2.2

7.4

46

FAST FASHION

There is no feasible way that your t-shirt should cost less than a sandwich and a cup of coffee. It just *shouldn't*. Fast fashion and the textile industry is wreaking major havoc on the global environment. The cost of a cheeky new shirt or cute pair of jeans may cost you next to nothing, but cut-price fashion is costing the Earth *everything*.

The majority of fast fashion clothes are made in countries where labour laws are lax. Most sweatshops are found in Asia, Eastern Europe, and Central and South America – and it is the sickeningly cheap labour in these sweatshops that allow wealthy western countries access to such cheap fast fashion. Bangladesh is the world's biggest exporter of fast fashion with over 4.5 million people employed in the garment industry. Employees, the majority of which are women, are often poor and subjected to terrible working conditions. These garment industry workers are paid a pittance for their labour, the average wage being about US$68 per month. Once manufactured, items are shipped vast distances by freight, truck or train to wealthier nations, which further contributes to the already enormous carbon footprint of fast fashion.

In the past 25 years, this model of cheap labour and exploitation has grown exponentially. Between 1994 and 2004, textile production increased by 400% to 80 billion garments per year. With fast fashion outsourcing production in this way – selling clothes for cheap but still with a big enough mark-up to turn a profit – huge amounts of waste is created. This **take, make, and waste** model also causes patterns of exploitative human labour practices with large corporations damaging the health of farmers and textile workers while increasing inequality and modern-day slavery.

As touched on, the environmental cost is as enormous as the human cost. The textile industry is the second most polluting industry after oil. Yup. Big, bad oil! Many oil derivatives are used to make synthetic fibres such as polyester and nylon, and petrochemicals are used in the treatment and dying of fibre, yarns and textiles. When we buy, wear and then wash clothes made of synthetic fibres they release plastic microfibres which get into the water supply. Oh, and let's not forget the use of agrochemicals, herbicides, insecticides and artificial fertiliser used for the monoculture farming of textile crops such as cotton. The carbon footprint (although we hate that term) also extends to shipping (textile products and clothing), packaging for shipping, packaging in shops, the running of the factories, and many other overheads, all combining to make this one of the most damaging and wasteful aspects of modern life.

SOME ENVIRONMENTAL IMPACTS OF THE FASHION INDUSTRY

1.5 trillion litres of water are used in clothing production

Produces 10% of global carbon emissions

Deforestation

70 million barrels of oil used each year to make polyester

Only 15% of garments are recycled

Water pollution from untreated waste-waters

Soil degradation from over grazing and chemical fertilisers

Water pollution from pesticides and fertilisers

Microfibres are released into water when you wash

YOUR COTTON T-SHIRT

Ah, the cotton t-shirt. A staple in everyone's wardrobe.

It is also one of the worst things to ever happen to the environment.
Globally, we consume between one and two billion t-shirts a year, and
cotton t-shirts are one of the most common garments in the world. The
t-shirt is also a complete ecological and human rights nightmare.

A typical cotton t-shirt begins its life in North America, Brazil, China or
India. Cotton farming requires planting, irrigation and watering in order to
harvest the tiny balls of fluff that surround a cotton seed. A staggering 2,700
litres of water are needed to grow enough cotton to make an average t-shirt
– that's more than the average human drinks in two and a half years!

The cotton industry also uses more pesticides and insecticides than
any other crop. These pollutants are harmful, some are carcinogenic
(cancer-causing) and put farm workers and surrounding ecosystems at
huge risk. Sure, organic cotton does not use pesticides or insecticides...
but it makes up less than 1% of the 22,700,000 metric tonnes of cotton
produced globally every year.

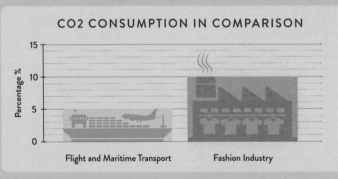

CO2 CONSUMPTION IN COMPARISON

Percentage %

15

10

5

0

Flight and Maritime Transport   Fashion Industry

Once the cotton is collected and processed into bales, it is shipped around the world to industrial mills in India and China where machines are used to blend, card, comb, pull, stretch and finally spin the cotton into yarn. In the case of t-shirts the fabric is knitted in a big machine, and the yarn is then processed in looms that weave or knit the yarn into fabrics. These fabrics are then bleached, chemically treated and dyed. Some of the dyes used contain cadmium, chromium, lead and mercury – all of which are carcinogenic (and yes these also come from mines, which cause their own environmental problems). Contamination from the chemicals in textile processing can leach into water systems, poisoning communities and ecosystems local to these mills. Toxic waste-water from these mills contaminates rivers and oceans. A recent example of this is the village of Jenggot, Indonesia. After a flood hit a local dye factory in early 2021, the village was submerged in possibly deeply unhealthy blood-red water, and videos taken by locals caused outrage on social media.

Once dyed and handled, finished textiles will then be shipped to factories in Bangladesh, Turkey, Vietnam, China and India where human labour is often exploited to make them into your t-shirts. There are clothing factories all over the world – they often independently supply companies such as major fashion labels and high-street brands, which conveniently gives big brands the distance to say that they "didn't know" about the human exploitation that occurs in them, and that their production was being outsourced. Yeah, *right*.

The finished t-shirt will then travel through warehouses, via trucks and ships (further adding to emissions), and is likely to be wrapped in plastics and packaging that is non-recyclable or simply dumped as part of the process. Until *finally* it finds its way to your high-street or internet shopping basket.

BREAKING UP WITH FAST FASHION

Saying goodbye to fast fashion is the hottest trend there is.

Breaking up with fast fashion starts in your wardrobe, and on your shelves, and on that chair where the clothes are not quite dirty enough for the laundry but haven't yet made it back to the drawer. Sustainability and style don't have to be enemies: there are loads of ways to stay looking sharp while saving the planet.

Firstly, consider shopping second-hand or vintage. It's not just charity shops that stock hidden gems, but second-hand clothing has gone online. Apps and sites like Depop, Vinted and Vestiaire Collective are just some of the platforms where you can buy and sell second-hand and vintage clothes. These are also great platforms for selling your own unwanted clothing, so **don't** throw clothes away when you wish to get rid of them. This is a whole other waste problem: give them away to friends, donate to charity shops, recycle, sell, personalise, upcycle, or re-appropriate as cleaning rags or for craft projects. So many options!!

If you do need new clothing that you can't get second-hand – we're not advising you to buy underwear second-hand – look at the label of your clothes. Try to buy traceable, sustainable and ethically-produced clothes where makers have been paid well and treated fairly. Look for items that are made from recycled cotton, wool or other fibres, and try to shop organically if you can afford it. Eco-textiles are another solution; they can be made from fibres or materials sourced from agricultural waste such as stalks and leaves, algae or mycelium.

Last but not least, look at how you wash! The average US household does 400 loads of laundry per year, using a whopping 16,000 gallons of water per year. Washers and dryers both at home and in laundromats require huge amounts of energy. Dryers in particular use four to six times more energy than washing machines. Change the setting on your washing machine to wash clothes at a lower temperature and air or line dry clothing if possible to save the energy used in drying. Another tip is to try spot-washing and washing clothes less frequently.

Green, my darlings, is the new black.

"BUY LESS. CHOOSE WELL. MAKE IT LAST."

Iconic fashion designer Vivienne Westwood
knows what she's talking about.

"FAST FASHION ISN'T FREE. SOMEONE, SOMEWHERE, IS PAYING."

British journalist Lucy Siegal has written extensively on fast fashion.

CHAPTER SIX

WATER

WHAT'S THE COMMOTION IN THE OCEAN?

What's happening in the ocean? And where the hell is Nemo?

The state of the oceans is really, *really* bad. Climate change is causing the ocean to warm, which is damaging whole ecosystems whose inhabitants have evolved to only survive at specific temperatures. On top of simple heating, our oceans are facing a whole smorgasbord of crap right now, from overfishing to oxygen depletion, pollution from farming and industry, plastic and microplastic pollution as well as, to top it all off, poaching of creatures such as turtles and whales. Oceans are facing a tidal wave of trouble.

About 72% of the planet is covered by ocean, and 90% of the world's water is in the oceans. The remaining 10% is freshwater that we as humans need to survive, proving how scarce and precious a resource clean water is. Harmful human activity has not only led to huge levels of pollution in the oceans, but we have already messed it up so badly that we have transformed the actual chemistry of seawater. Increases in CO_2 have caused the acidity of the water to increase. This increase in acidity makes it hard for shelled marine molluscs and corals to grow. Ocean acidification has other less obvious affects – such as to the sense of smell of many fish, damaging their ability to navigate – and changes how some fish behave – making them unable to escape predators. Jellyfish appear to be thriving in the changing seas, but not much else. Never trust jellyfish, the slippery buggers – they be beautiful but they be painful.

CORAL REEF BLEACHING

Coral reefs are incredibly biodiverse places. They cover less than 1% of the ocean floor yet support an estimated 25% of all known marine species. Not only do they support intricate life systems, but coral reefs also capture carbon. But coral reefs are under massive threat. As much as 27% of monitored reef formations have died, impacting rich and populous ecosystems. And 32% more are at risk of being lost within the next 32 years. Increasing temperature is the main cause of coral reef bleaching, as well as pollution runoff, ocean acidification, overexposure to sunlight and changing tides. Some corals are evolving adaptations to warming, but despite their efforts to survive it isn't enough for them to endure yet more stress.

While dead zones, hypoxia and coral reef loss are being observed all over the world, the 27-year decline of the Great Barrier Reef is probably the most well-known and depressing. Over half of the Great Barrier Reef is now dead due to climate change. Once the natural connections in an ecosystem are broken – as we have observed in nature already – it is very

······

"WATER AND AIR, THE TWO ESSENTIAL FLUIDS ON WHICH ALL LIFE DEPENDS, HAVE BECOME GLOBAL GARBAGE CANS."

Jacques-Yves Cousteau (1910–1997), was a French explorer, conservationist, filmmaker, innovator and scientist who extensively studied marine life. We doubt he'd be too happy with the state of the oceans today.

······

difficult for them to reconstruct themselves, and **impossible** for them to be fully restored or replaced. However, we can restore the oceans to some degree and some remote and protected reefs in the Pacific are beginning to recover.

OCEAN POLLUTION

According to the World Wildlife Fund, 80% of the plastic produced on land ends up in the ocean. Of this amount, 90% of plastic waste enters the world's oceans through ten major rivers in Africa and Asia. But where is all this plastic coming from?

Even the term plastic covers a whole range of materials. Essentially, the term plastic refers to a material that is easily moulded into different shapes (so even cheese could be a plastic). It is considered a synthetic material which can be created using a wide range of polymers (materials made of long, repeating chains of molecules) such as polyethylene, PVC, nylon, etc. These polymers, the basis of all plastics, come from natural materials such as cellulose, coal, natural gas, salt and crude oil. These can be shaped and reshaped into multiple forms while soft, and they can then be set to become rigid or slightly elastic.

This is where our old enemies, oil companies, make an unwelcome reappearance. Oil companies make an enormous amount of money from supplying crude oil for the purposes of creating plastics. In fact, the invention of single-use plastics was their idea. After an explosion of plastic use in the 1960s, the oceans have suffered obscene amounts of pollution. Today over one million plastic bottles alone are used and thrown away every minute. Plastic straws are the third most common type of waste found on beaches, with cigarette butts taking pole position for pollution.

HOW LONG PLASTICS TAKE TO DEGRADE IN THE OCEAN

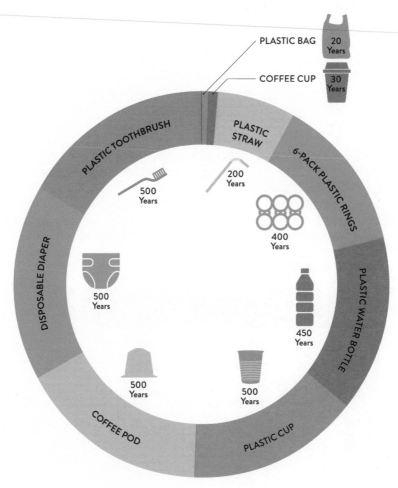

PLASTIC BAG — 20 Years

COFFEE CUP — 30 Years

PLASTIC TOOTHBRUSH
500 Years

PLASTIC STRAW
200 Years

6-PACK PLASTIC RINGS
400 Years

PLASTIC WATER BOTTLE
450 Years

DISPOSABLE DIAPER
500 Years

PLASTIC CUP
500 Years

COFFEE POD
500 Years

HUMAN WASTE

One of the worst contaminants of our oceans is poop. Yep, you heard that right, and worse... it's human poop. A WHO report from 2017 stated that 1.4 billion people have access to only basic water services and 2 billion people do not have access to facilities such as toilets or latrines at all. It is the crappiest situation in the world. In America alone, 1.2 trillion gallons of untreated sewage, storm water and industrial waste are discharged annually.

Not only does a lack of proper sewerage infrastructure pollute the ocean but it also seriously hurts vulnerable communities. About two billion people worldwide still lack access to clean water and regular waste collection, allowing contaminated water to spread diseases such as cholera and typhus. Annually, 829,000 people die due to diarrhoea caused by unclean water, including 297,000 children. Proper sewerage saves lives.

According to the WHO:

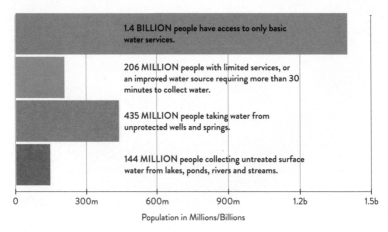

1.4 BILLION people have access to only basic water services.

206 MILLION people with limited services, or an improved water source requiring more than 30 minutes to collect water.

435 MILLION people taking water from unprotected wells and springs.

144 MILLION people collecting untreated surface water from lakes, ponds, rivers and streams.

| 0 | 300m | 600m | 900m | 1.2b | 1.5b |

Population in Millions/Billions

FRESHWATER PROBLEMS

It's not just the ocean that's affected by toxic pollution. Rivers, lakes and freshwater sources are also suffering. Temperature increases caused by climate change are affecting the flow of water into lakes and preventing the mixing of oxygenated and deoxygenated water, damaging the health of aquatic organisms.

Freshwater pollution is a global problem. In the USA 40% of rivers and 46% of the lakes are polluted and unsuitable for swimming, fishing or any other activity. Likewise, the Asian continent has a very high number of contaminated rivers. The Ganges river in India is considered the most polluted river in the world, while groundwater in Bangladesh is contaminated with harmful carcinogenic industrial chemicals including arsenic and mercury. In China, a 2018 report found that up to 86.2% of China's groundwater is of "bad" or "very bad" quality. Safe drinking water systems and safe sanitation that effectively disposes of human waste are essential to ensure cities and towns grow in a way that is both healthy and sustainable.

While pollution of water sources is a major issue, so is the lack of clean water sources caused by climate change. Cape Town and other major cities have experienced extensive drought and urbanisation resulting in an ongoing water crisis. In fact in 2017–18 Cape Town became the fist major global city to announce an apocalyptic countdown to "Day Zero", when the city would actually **run out of water**. However, with cooperation and understanding of the severity of this problem, people listened to strict water restrictions and reduced their household use of water to help curb the problem. Despite these measures, Cape Town continues to experience water problems, but it is no longer at risk of a major fatal disaster.

THE GREAT PACIFIC GARBAGE PATCH

An ocean gyre is any large system of circulating ocean currents, particularly those involved with large wind movements. When combined with pollution, these currents catch and collect plastic waste in huge slow-moving circular eddies, causing massive garbage patches to form in the ocean. The biggest, the baddest and most despicable is the Great Pacific Garbage Patch (GPGP). It is located halfway between Hawaii and California and is currently 1.6 million square kilometres in size, three times the size of France, or twice the size of Texas. It can even be seen from space, it's that big!

Scientists estimate that 1.15 to 2.41 million tonnes of plastic enter the ocean each year. Once these plastics get caught up in the gyre, they are unlikely to leave the garbage patch until they degrade into smaller microplastics under the effects of sun, waves and marine life (we'll cover microplastics overleaf). As more and more plastics are dumped into waterways, the microplastic concentration in the Great Pacific Garbage Patch will only continue to expand its already devastating toll.

According to calculations by scientists who study the patch, the problem is even worse than they first anticipated. According to a 2018 report, there were more than 1.8 trillion pieces of plastic in the patch and that number is growing. That's right, not one billion... *one trillion*. This is the equivalent of 250 pieces of debris for every single human in the entire world. The mass of the plastic in the GPGP was estimated to be approximately 80,000 tonnes, the equivalent of 500 Jumbo Jets. And the patch is only getting bigger; since measurements began in the 1970s, calculations have shown that the microplastic concentration has increased exponentially, proving that there is more and more debris arriving each day. Unless we do something about it, it's only going to get worse and worse.

Animals have suffered greatly around the garbage patch. Due to the size and colour of plastic waste, animals confuse it for food, causing malnutrition as they fill their stomachs on useless junk. For example, turtles tend to confuse plastic bags for jellyfish, their usual food source. Fishing nets and similar debris account for 46% of the GPGP – these "ghost nets" entrap marine life, which, unless a friendly human happens to be floating by (unlikely in a massive garbage patch) then dies. The GPGP is threatening the very existence of Pacific marine life. Studies have shown that about 700 species have encountered the marine debris, and 92% of these interactions are with plastic. Of the species affected by plastic, 17% are already on the IUCN (International Union for Conservation of Nature) Red List of Threatened Species.

Even for the animals that avoid entanglement in ghost nets, the passing of plastic up the food chain is an area of serious concern. Through a process called bioaccumulation, chemicals in plastics will build up in the body of the animal feeding on the plastic, and as the feeder becomes prey, the chemicals will pass to the predator – making their way up the food web. These microplastics can be found in humans as well.

Plastic pollution costs us too much. You cannot put a price on healthy and clean oceans, but according to a study done by the Ocean Clean Up and Deloitte, yearly economic costs due to marine plastic are estimated to be between $6–19 billion. These costs stem from its impact on tourism, fisheries and aquaculture, as well as (governmental) clean-ups. While these costs attempt to include the impact on human health and the marine ecosystem, only time will tell of the true devastating cost of the Great Pacific Garbage Patch.

MICROPLASTICS

If you thought bigger pieces of plastics were causing mass destruction to the ocean, just wait till you hear about microplastics. Because this crap tornado is about to get a whole lot worse.

So, you use a plastic bag and throw it away. What happens next? You could recycle it, that would be great! But 90% of the world's plastics are not recycled, so what happens to them? The average plastic bag is only utilised for 12 minutes, but takes approximately 20 years to break down in the ocean, and even then it is not gone completely. The discarded bag slowly erodes, breaking down into smaller and smaller pieces known as microplastics. Microplastics are not a specific kind of plastic but are any type of plastic fragment that is less than 5mm in length, according to the US National Oceanic and Atmospheric Administration (NOAA). Microplastics badly pollute the environment – leaching into soil, contaminating water and embedding themselves into the bodies of living organisms across the globe.

Humans use a million plastic bottles a minute and plastic straws are non-recyclable because of their size. The same properties that have made plastics so ubiquitous in modern life – their waterproofing, their durability, their ability to be moulded – are the exact same things that are causing them to choke up our oceans and harm part of the marine food web. We need to lobby for changes in waste management, as well as rethinking how we use and treat materials. Some have adopted bioplastics as the way forward. Bioplastics are plastics made from renewable biomass sources, such as vegetable fats, lactose proteins, oils and starches. Even straw, wood chips, sawdust and food waste can be transformed into bioplastics! Replacing single-use plastic in all areas of our lives is the first step, and the phasing out of petro-chemical plastic in favour of bioplastic can further help us on our way.

HOW PLASTICS ARE CONSUMED BY MARINE LIFE

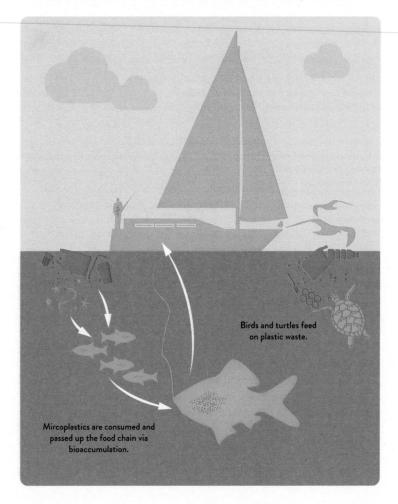

Birds and turtles feed
on plastic waste.

Mircoplastics are consumed and
passed up the food chain via
bioaccumulation.

BIOACCUMULATION IN POLAR BEARS

We love polar bears. We really do. But it's understandable if they don't feel the same way about us. I mean we're really screwing them over right now. As smaller marine life at the bottom of the food chain confuse microplastics for food, this causes a build-up of toxic chemicals in their bodies, before they're consmued by predators. This process of slow and steady poisoning is called bioaccumulation. Once ingested, certain types of molecules are more likely to remain within an organism. Toxic industrial and agricultural chemicals have this property, particularly those that are lipophilic (attracted to fat molecules). Heavy metals, radioisotopes, insecticides, herbicides and other poisonous compounds also play a part in this problem.

Once a small organism, such as an insect or a fish, is eaten by a bird or a predator, the toxins build up even further in the next part of the food chain. That predator can then be eaten by another, bigger predator, and the toxins get passed along again. This is called biomagnification, which eventually leads to our polar bears.

Apex predators, like the fluffy polar bear, are most at risk of biomagnification, where the concentration of toxins is magnified due to the ingestion of lots of other species who have microplastics or similar in their systems. This results in a gradual poisoning that can have all sorts of horrible effects. Studies have shown polar bears are having many developmental issues because of the poisons they are exposed to, including declining skull density. Not only that, these toxins are causing male polar bears to have weaker penis bones (it's a thing!), meaning that they can't get it on and make baby polar bears, causing a species-wide epidemic of sheepish apologies and faked headaches.

BIOMAGNIFICATION

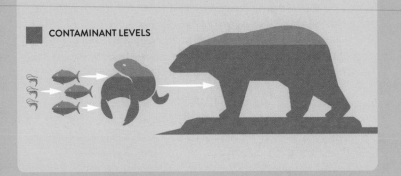

CONTAMINANT LEVELS

"AS GOVERNOR OF ALASKA, I TOOK
A STAND AGAINST POLITICIZED
SCIENCE WHEN I SUED THE FEDERAL
GOVERNMENT OVER ITS DECISION TO
LIST THE POLAR BEAR AS AN ENDANGERED
SPECIES DESPITE THE FACT THAT
THE POLAR BEAR POPULATION HAD
MORE THAN DOUBLED."

Republican Governor Sarah Palin clearly
hates innocent polar bears. It will also
be no surprise whatsoever that the
population of polar bears had not, in fact,
doubled.

TACKLING POLLUTION

As we've seen, the damage caused by plastics in the ocean is extensive, so to keep it from getting even worse we need to stop plastic entering the oceans in the first place. We need to deal with waste at its source more efficiently on individual and community levels, by actions such as removing plastic from rivers before they meet the sea.

INTERNATIONAL CHANGE

We are just beginning to see national initiatives and commitment beyond individuals giving up single-use plastics. Rwanda, Kenya and Chile have just banned plastic bags, and Costa Rica aims to do the same. Most countries in Europe, including the UK, have introduced a levy on plastic bags as a deterrent to consumers and supermarkets.

Another international organisation committed to tackling ocean pollution is the Clean Oceans Initiative. Launched in 2018 by the European Investment Bank, it is aimed at supporting coastal communities in Asia, North Africa and South America. The initiative offers expertise and up to €2 billion over five years to help fund projects that manage plastics and pollutants at their sources and aim to clean up waste-water before it reaches the ocean. Other examples of projects include a €50 million loan for Kotoneau on the South Coast of Benin for a storm-water renovation project. This is projected to help 187,000 people and cut down on pollutants before they reach the Gulf of Guinea. Another €80 million loan has been given to Buenos Aires for water sanitation and services. This will include the upgrade and extension of pre-existing sewerage and water treatment systems and create access to safe water for over 20,000 people. There are more projects like this happening in various places around the world.

While initiatives like these support amazing projects, critics note that funding is in the form of loans to already indebted and developing countries. Supporting sustainability should be interest-free at the very least.

NATIONAL CHANGE

Community and national level coastal clean-ups are important. While it may seem like their total impact is tiny, they strengthen communities, raise awareness and help nurture our connection to the sea. Dealing with plastic pollution and waste in coastal regions (before it gets a chance to really get into the ocean) is a key strategy. Did you know over ONE TONNE of plastic waste enters the ocean every four seconds? By paying more attention to what happens to our waste we can really start lobbying for change.

On a national governmental level, industries that keep pushing the use of single-use plastic should be pressured to change to non-plastic alternatives. One way of doing this is taxing the use of non-essential plastic in industry, and subsidising biodegradable packaging alternatives. With a population of eight billion, it is not possible for us to have 100% pristine oceans ever again, but even 20% closer to pristine is better than zero.

PERSONAL CHANGE

Our daily choices really do make a big difference! If you are financially able to, buy from sustainable retailers who can tell you all about how the fish they sell are being caught or farmed. Try to avoid endangered species like tuna, salmon and swordfish and reduce your fish consumption if you can.

Cut down your own water use by retro-fitting taps with adapters that diffuse the flow of water. Be respectful and considerate when you visit the seaside, river or lakes, whether that be in your home country or on holiday.

Swap out regular sunscreens for mineral versions. Sunscreens currently contain chemicals which are endocrine disruptors and harmful to the development of coral reefs. The Hawaii Sunscreen Act comes into effect in 2021 and bans the use of sunscreens that contain oxybenzone and octinoxate, which are chemicals that contribute to coral bleaching when they wash off us and into the ocean.

And last, but by no means least, reduce your use of single-use plastic where possible. Carry a keep cup instead of getting take away coffee, use cloth bags or tote bags while shopping. These may seem like small changes, but together we can make a real difference.

OVERFISHING

What if there aren't plenty more fish in the sea?

More than 90% of all the world's large fish have been lost since the 1950s. From the remaining 10% that exist today, 90% are either fished to capacity or overfished. Fishing companies are fishing further from shore and deeper into the ocean, using military technology like radar and helicopters in order to find the dwindling remains of already scarce populations.

Only 1% of previous populations of Atlantic salmon survive in the present day. Marine predators such as sharks, tuna and trout are also declining in numbers. Because they are apex predators this is having huge knock-on effects on the ocean's ecosystems.

OVERFISHING: THE FACTS

Fisheries could be worth an
EXTRA US$50 BILLION
every year if managed sustainably.

3 BILLION PEOPLE
rely on fish as their
primary source of protein.

75% of global fisheries
are under-performing.

Fisheries
contribute
US$274 BILLION
a year to
global GDP.

**260 MILLION
PEOPLE GLOBALLY**
are employed directly or
indirectly in fishing,
97% of these are in
developing countries.

The global fish
harvest could be
40% HIGHER
if under sustainable
management.

The value of the
Pacific Halibut fishery
has **INCREASED BY 222%**
since the introduction of sustainable
management measures.

How we treat marine life needs to change on both an industrial and retail level. Lobbying for waste reductions in supermarkets and for changes in industrial practice can start the process off swimmingly. Redesigning and retrofitting fishing gear can help reduce by-catch. New fish traps that only catch the desired fish and can release 80% of the by-catch and wouldn't affect fishers' incomes. Pre-existing fish traps can be retrofitted with escape traps, a very low-tech and cheap solution to waste.

MARINE BIODIVERSITY EXTINCTIONS

Fishing is having a devastating impact on life in the oceans and having an equally devastating effect on the balance of marine life. Trawlers remove adult fish before they have the time to breed, and rather than fishing a smaller amount of fish over a long period of time the fishing industry is choosing to remove large amounts of fish before populations have time to recover. In addition to this, fishing quotas and economic difficulties make it hard for fishing communities to do much else except keep pace with these unsustainable practices.

Wetlands and mangrove forests are being destroyed to make way for development. Coastal habitats and ecosystems are some of the most diverse in the world and also act as the natural filtration systems between land and the sea. These habitats are the nurseries of the oceans, and destroying them disrupts the life cycles of many fish and sea creatures, causing some to become extinct.

PRAWN COCKTAIL FROM HELL

If there's any fish that causes environmental concern, it's the humble prawn (or shrimp if you want to get all American about it). The majority of prawn fishing is horrifically unsustainable as they are caught by gigantic nets that are weighed down and dragged along the ocean floor, bulldozing entire habitats and capturing anything in its path. Moreover, many fish farms are constructed in places with rich pre-existing ecosystems, which become irreparably damaged by the farming practices. In the USA, for example, they are often built where mangroves and wetlands used to exist.

The farms are created when artificial ponds are netted off from the rest of the water in order to farm shrimp (and other fish, too). These farms are forced to use antibiotics to treat fish that get sick from the overcrowding in these nets, which are themselves harmful pollutants. Moreover, there is often a hidden human cost to shrimp on a stick. There is an increasing body of evidence that many shrimp processing factories in South-East Asia use slave labour to peel and process shrimp in order to meet ever-increasing market demand. Without proper regulation in future, fish farming will continue to cause vast harm to both ecosystems and coastal communities.

Financial issues make this even harder to solve. The price of fish can be very deceiving, Tuna, for example, is a relatively scarce fish, but tuna fishing is so heavily subsidised that it appears to be a cheap food. Regardless of this, fishing communities still tend to be poor as the margins are so tight. Seafood that is genuinely sustainable does exist, but it is rare and expensive. If you can, try to ensure that the prawns you buy are farmed in a manner approved by marine stewardship and protection programmes. Many shrimp farms in Northern Europe are monitored by marine stewardship programmes. These shrimp, however, are much more expensive, making them less accessible to the average consumer.

SUSTAINABLE FISHING;
SAVING NEMO

Poor Nemo, he's screwed whether he's found or not. Unless, of course, we start doing something to help the oceans.

Iconic oceanologist and marine biologist Dr Ayana Elizabeth Johnson explains that part of the challenge is that "some fishing traditions do not scale up, fishing for a global population of one billion people is not the same as fishing for eight billion." Simply put, as we approach a global population of eight billion people, we cannot keep fishing in the same way. It is difficult for communities who have strong traditions tied to the ocean with growing populations to maintain some of the traditions and continue to use the practices we use today. Johnson goes on to say that coastal communities (many of which are often poor fishing communities) need to be engaged and empowered as part of the solution, instead of forcing them to struggle in vain to meet the demands of the market.

Subsidies for fishing, for instance, could be used to reinvest in and support coastal communities to prioritise sustainable fishing methods and conservation, rather than simply using them to decrease the final price for the consumer in an unsustainable manner. How people are using the oceans and the social and economic problems coastal communities experience are intertwined, and must be addressed together when thinking about how to solve the problems the oceans face.

There is also evidence that changes to fishing practices can improve fish populations in specific areas. This suggests that marine ecosystems could recover in areas that are protected by law. Some preserved marine ecosystems are already showing signs of recovery – Monterey Bay in California is one successful example. Establishing protected areas can keep fish populations safe and can benefit the community by leading to a more sustainable form of tourism. Work can be done with hotels, tourist attractions in coastal communities and local businesses to change the ways in which they dispose of their waste.

Individually, each of us can make daily choices to reduce the harmful impact of overfishing. Sardines and other small fish, such as anchovies, are more sustainable to farm because they reproduce quickly – as opposed to salmon and **avoid tuna, shark and swordfish** – as top predators, they are the equivalent of bears and tigers in an aquatic environment habitat. They take a long time to reach maturity, so are harder to replace when they are gone.

Do your bit for poor, cute Nemo!

"THE OCEAN IS LIKE A CHECKING ACCOUNT WHERE EVERYBODY WITHDRAWS AND NO ONE MAKES A DEPOSIT. THAT'S WHAT'S HAPPENING BECAUSE OF OVERFISHING."

Enric Sala, a former university professor who was so distressed at what was happening to our oceans that he quit academia to become a full-time conservationist.

CHAPTER SEVEN

HUMAN HABITATS

URBANISATION

According to the UN, approximately four billion people live in urban centres. That's over half of the world's population. This percentage is only increasing: by 2050 it's predicted that two-thirds of the world's population will live in cities. How a city's resources are managed, how they are distributed and who benefit from them are far from equal. Urban centres magnify the inequality in our societies, and in the world as a whole. In fact, just under one-in-three of those in urban centres live in slum conditions. But what does this have to do with climate change? Simply put, pretty much everything.

High levels of air pollution take a major toll on public health in cities all over the world. Washington, D.C., Athens, London, Shanghai, Paris and Mumbai – all experience major issues concerning air pollution. A study by the Health Effects Institute found that air pollution led to around 852,000 premature deaths in China alone in 2017.

You might think that by staying indoors, especially on smoggy days, you can protect yourself, but even if you are fortunate enough to be able to lead a lifestyle that allows this, is not just in the street that air pollution is an issue, without expensive filtration systems it permeates our homes, too. Unfortunately, it can be difficult to monitor indoor air quality, and this is exacerbated in poorer households which are more likely to depend on burning solid fuels. In China – in 2017 again – indoor air pollution resulted in an additional 271,100 deaths.

THE SOURCES OF URBAN AIR POLLUTION

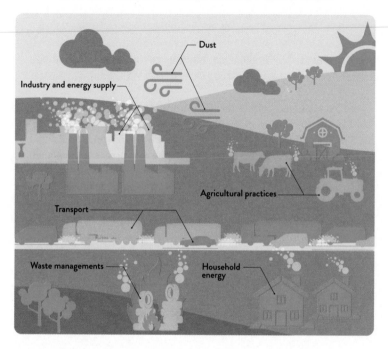

Based on a 2018 report, the top ten most air-polluted cities were all located in the Global South:

1. Ghaziabad, India
2. Hotan, China
3. Gujranwala, Pakistan
4. Faisalabad, Pakistan
5. Delhi, India

6. Noida, India
7. Gurugram, India
8. Raiwind, Pakistan
9. Greater Noida, India
10. Bandhwari, India

Air pollution is a big deal. It is literally life or death (see the infographic below). Non-communicable diseases (those that cannot be passed from one human to another) account for 72% of deaths globally. The World Health Organisation (WHO) identifies an average of 4.2 million deaths every year caused by ambient air pollution. Of the world's population, 91% live in areas where air quality exceeds safe WHO limits.

As the damage from pollution has become more apparent, more countries are looking to green alternatives to prevent further damage to the Earth. Solar and wind energy, eco-friendly building materials, and non-toxic products are increasingly being used to preserve the planet. While these green initiatives are taking place around the world, some countries have a long way to go, and it is most commonly those in less developed countries, who have the highest levels of air pollution, that are least able to spend the money on these sorts of policies.

AIR POLLUTION DEATHS

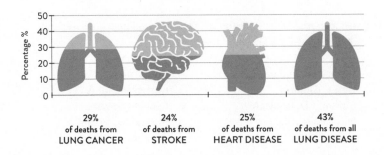

29%
of deaths from
LUNG CANCER

24%
of deaths from
STROKE

25%
of deaths from
HEART DISEASE

43%
of deaths from all
LUNG DISEASE

CANCER ALLEY

Even within more wealthy countries, those most likely to be affected by air pollution are often marginalised or low-income communities. One example of this is "cancer alley" in Louisiana. Residents in this area face the highest risks of cancer and respiratory diseases from air pollutants in the USA, mainly resulting from the industrial activity that occurs along the banks of the Mississippi. In the riverside community of Reserve, the risk of cancer is the highest in the USA, reaching 50 times the national average.

The residents of this community are predominantly poor, Black and working class – the populations most likely to be disproportionately affected by pollution. Poor communities do not have the resources to move to less polluted areas, but even so some people along Cancer Alley have been forced to abandon their homes.

"MODERN AIR IS A LITTLE TOO CLEAN FOR OPTIMUM HEALTH."

DENIER
DENIED
DENIER

Robert Phalen, an "air pollution researcher" raised eyebrows in 2012 by denying that air pollution was a risk to children. Was it any surprise, then, when Trump appointed him to a key USA EPA advisory committee?!

SUSTAINABLE CITIES

If cities are where most of us live and where most of us are likely to be affected by climate change, surely we should be doing our darnedest to make sure they're liveable, at the very least? As well as being centres of air pollution, cities are more vulnerable to flooding, overheating and irregular temperatures, all of which are threats to life and damaging to the economy. Cities are the most densely populated human communities on the planet, so designing and developing cities to be more resilient to climate change is an important priority across the world.

EXAMPLES OF SUSTAINABLE CITIES

Athens, Greece is a dense urban environment that is about 10°C hotter than suburban areas due to the density of human activity and industry. This is called the "Heat Island Effect", and many cities are taking on large

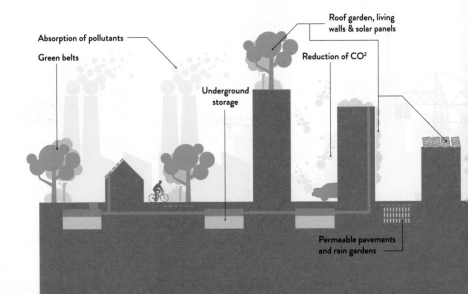

Absorption of pollutants

Green belts

Roof garden, living walls & solar panels

Reduction of CO²

Underground storage

Permeable pavements and rain gardens

scale urban-planning schemes to tackle it. Athens has begun focusing on conservation and plans to use nature-based solutions to reduce the impact of the Heat Island Effect, such as creating 25% more green spaces and creating green corridors for species and air to move.

In 2016 a state of emergency was announced in Mexico City due to catastrophic levels of air pollution. In response, the Via Verde is an attempt to introduce plants in vertical gardens of ivy, foxtail and aralia (a hardy, tough-as-nails species) at the sides of polluted roadsides, which will be watered using reclaimed rainwater rather than drinking water. This nature-based solution aims to filter 27,000 tonnes of exhaust and remove 5,000 kilograms of dust from the air every year.

SUSTAINABLE CITIES

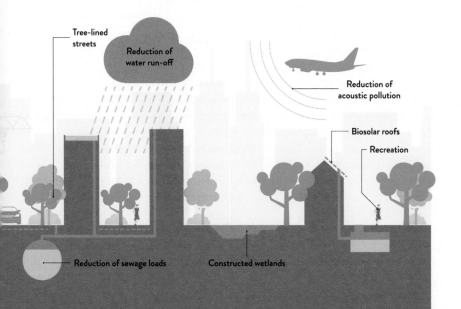

Tree-lined streets

Reduction of water run-off

Reduction of acoustic pollution

Biosolar roofs

Recreation

Reduction of sewage loads

Constructed wetlands

ENVIRONMENTAL RACISM

Not only is climate change pretty much the worst thing that can happen to humanity... it's also racist. Yup, as if it could get any worse, this is an actual thing and unsurprisingly it's called environmental racism. This refers to the ways in which waste, pollution and the climate crisis in general disproportionately impact Black people, Indigenous people and people of colour. The term was first coined in 1980 by Dr Benjamin Chavis, who was researching the correlation between racial demographics and toxic waste locations, and is one of the many reasons why racism is a public health crisis. "Racism is what has made it possible to systematically look away from climate threats for more than two decades," he wrote. "It is also what has allowed the worst health impacts of digging up, processing and burning fossil fuels – from cancer clusters to asthma – to be systematically dumped on indigenous communities and on the neighbourhoods where people of colour live, work and play."

Environmental racism is a global issue and in many ways a hangover of colonialism. The USA is a microcosm of some of these global issues around unfair racial divides. Of people living within a 3km range of hazardous waste from industrial plants or urban environments, 55% are people of colour (*far* more than you would expect naturally from the overall population demographics of the USA), and the remaining 45% are invariably poor. Of those living close to hazardous chemical facilities, 47% are Black or Latino. Black communities are exposed to 50% more air pollution than white communities and communities with more people of colour are 40% more likely to be drinking from water sources that are unsafe.

CLIMATE CHANGE AND BLACK LIVES MATTER

Racial justice is vital to saving the planet. During the summer of 2020, the world saw people from all over the world march in protest against the murders of George Floyd, Breonna Taylor and many more at the hands of police. A powerful movement with multiple leaders, there is a lot that can be learnt from the Black Lives Matter organisation that would benefit the work and results of the climate movement.

Challenging white supremacy, and its social, political and economic reach, is perhaps the biggest challenge to protecting life and the planet. Both racism and climate change are systemically, historically, materially and geographically entwined. Systemic problems require systemic solutions, and existence is intersectional. Every part of life is somehow linked to every other part, and we can't tackle these issues in isolation. As Dr Ayana Johnson states in her How To Save A Planet podcast: "if we don't work on both [racism and climate change], we will not succeed".

FLINT, MICHIGAN

The town of Flint, Michigan, the centre of a global water pollution scandal, is an example of how racism worsens public health issues. Before 2014, the water source used by the city of Flint, Michigan was from the Lake Hudson and the Detroit River, which was treated and managed by the Detroit Water and Sewerage Department. However, in 2014, this source was changed to the Flint River. The water in the Flint River, however, was not treated properly and dangerous metals, such as lead, started leaking from old pipes into the water supply. Elevated levels of lead and other heavy metals started to contaminate the water supply, exposing over 100,000, **majority Black,** residents to increased concentrations of lead.

PERCENTAGE OF CHILDREN IN THE USA (AGED 1 TO 5) WITH HIGH BLOOD LEAD LEVELS (2007-10)

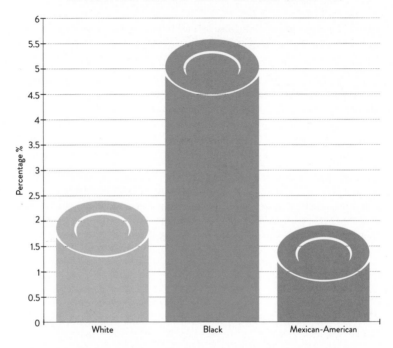

In January 2016, a state of emergency was declared due to data that showed that between 6,000 and 12,000 children were exposed to drinking water containing high levels of lead. As well as this, inadequate treatment of the water caused outbreaks of Legionnaires' disease that killed 12 people and affected 87 others. Fifteen criminal cases have been filed against state officials, but to date only one conviction has been obtained while the rest were all dropped.

INTERSECTIONAL ACTIVISM

Reporting on how climate movements in the west have catered quite exclusively to white middle class people is something of an issue. Anyone that cares about climate change and is willing to put in the hours in the fight against it is, of course, very good! But to not actively include Black, Indigenous, East, South and Central Asian communities is hugely problematic – after all these are the communities who already bear the brunt of climate change, and those who can least afford to change their lifestyles to combat it without causing increased levels of poverty. They need to be given the space to be at the forefront of the movements.

What we can do:
- Centre Black and Indigenous voices in the discourse concerning climate change.
- Support and protect Black and Indigenous communities.
- Listen to and *believe* what Black people say about their experience.

"YOU'RE NEVER TOO YOUNG OR TOO SMALL TO CHANGE THE WORLD."

Mari Copeny, Flint resident, who was eight years old when she wrote to then-President Barack Obama asking for help with her town's water supply. She is now a youth activist.

STANDING ROCK PROTESTS

"There is a long history in the Americas of beautiful pieces of wilderness being turned into conservation parks – and then that designation being used to prevent Indigenous people from accessing their ancestral territories to hunt and fish, or simply to live."
–Naomi Klein, *Let Them Drown*

Indigenous communities have long borne the brunt of colonialist greed. Many Indigenous communities have been restricted to reserves for generations, and the legislation for the protection of the national parks in the USA, Australia, New Zealand, Brazil, Colombia and Canada do not entitle Indigenous communities to live in their native lands the way they did before their lands were invaded by Europeans.

Deliberate pollution and disregard for the health of Indigenous people is a problem in every colonised country. Toxic contamination of water, land, air and resources from industrial agricultural pesticides and chemicals disproportionately impacts the health of Indigenous peoples. Additionally, the US and Canadian governments have consistently failed to provide funding to tribes and Alaskan villages to develop and implement environmental protection infrastructures. Polluting industries such as mining or oil refineries are often on or near Indigenous lands and governmental support regarding the safe management of solid and hazardous waste and wastewater is currently haphazard and inadequate.

THE DAKOTA ACCESS PIPELINE

The Dakota Access pipeline is a $3.7 billion project transporting crude oil from the Bakken oil field, North Dakota, to a refinery near Chicago. In 2016, it faced huge opposition from local Indigenous communities due to concerns over water contamination and destruction of sacred lands. The local Standing Rock Sioux tribe and thousands of Native American supporters from across North America set up camps in Cannon Ball to try and block the oil project. While the protesters continuously stated their peaceful intentions, they were met with force by the local police and national guard.

The tribe also fought the pipeline in the courts, saying that the project violated federal law and native treaties with the US government, as well as the United Nations' declaration on the rights of Indigenous peoples.

While Barack Obama did at least stall the pipeline, this was immediately overturned by Donald Trump. The pipeline has been active since the summer of 2017. A small oil spill has already occurred, and a large spill would affect access to clean water for the 17 million Indigenous Americans who live in the area.

Writing in *The Guardian* in 2019 Mike Faith, the chairman of the Standing Rock Sioux Tribe said that their fight isn't over. "To some, this may be just another pipeline in just another place," he wrote. "But to us, it's not just a pipeline, it's a threat. And it's not just a place, it's our home. The only one we have. Every day the pipeline operates represents a threat to our way of life and an insult to our culture and traditions that have withstood so much. We are still here. We are not giving up this fight."

CLIMATE POLICY

Here's the thing with climate change, as much as we can take responsibility on an individual level for our daily choices and actions, there's only so much one person can do. As well as this, what action looks like to different people varies widely depending on your level of financial security and privilege. Nowadays, privilege tends to get people's backs up, but privilege isn't all caviar and Range Rovers. Privilege is simply an indication of the material reality and lived experience of different people. We can have different types of privilege; financial, health, being able-bodied, gender and white privilege must be factored into how we design, plan and implement changes that make the world more liveable for everyone.

Not everyone can afford healthy organic food, free-range chicken eggs, organically raised ethically farmed meat, or sustainably sourced fish. Low income people cannot afford to buy the best quality and most ethically produced food to get the calories they need. Nor can they afford to buy essentials from sustainable companies. The privilege of choice isn't always available.

This is where climate policy comes in. We know that talking about policy can be daunting, and the conversation used is often clouded in complicated and political language. So, let's break it down and look at two of the most important international policy pieces of the last few years; the UN's Sustainable Development Goals and the Paris Climate Agreement.

SUSTAINABLE DEVELOPMENT GOALS (SDGS)

A sustainable future requires more than just reducing emissions – human issues must be addressed as well. Adopted by all United Nations Member States in 2015, SDGs address the global challenges we face, including poverty, inequality, climate change, environmental degradation, peace and justice. The Sustainable Development Goals, also known as the Global Goals, were adopted as a universal call to action to end poverty, protect the planet and ensure that all people enjoy peace and prosperity by 2030.

There are 17 SDGs comprising 169 targets, 3031 events and 5414 actions. These 17 goals are all intertwined, recognising that action in one area will affect outcomes in others, and that development needs to balance social, economic and environmental sustainability. They recognise that ending poverty and other deprivations must go hand-in-hand with strategies that improve health and education, reduce inequality, and spur economic growth – all while tackling climate change and working to preserve our oceans and forests. Part of the SDGs is the pledge to "Leave No One Behind" by fast-tracking countries that are furthest behind in terms of progress. Some of the most ambitious goals for the SDGs include several "zeros", including zero poverty, hunger, AIDS and discrimination against women and girls.

Everyone is needed to reach these ambitious targets, from businesses to governments to international non-governmental organisations. The SDGs show that climate change is not just one issue, but interlinked with poverty, inequality and corruption. We really need to get off our collective arse and work for it if we're going to reach that 2030 target.

THE PARIS CLIMATE AGREEMENT

On 12 December 2015, 196 parties (mainly countries) signed the Paris Agreement, a legally binding international treaty on climate change. The Paris Climate Agreement is a landmark declaration, being the first time a binding agreement brings all nations into a common cause to combat climate change and adapt to its effects. It entered into force on 4 November 2016.

The goal of the Paris Climate Agreement is to limit global warming to well below 2°C, preferably 1.5°C, compared to pre-industrial levels (remember the damage that 2°C can do? Flip to page 11 for a quick reminder).

However, researchers have predicted that there is only a 5% chance that the Earth will not warm past 2°C by 2100, and only a 1% chance that we will meet the more ambitious Paris Agreement goal of keeping warming below 1.5°C. There are issues to be solved every way you turn – even if all use of coal and oil stopped today there would still be an extra 0.7°C to 0.8°C worth of energy currently embedded in our climate. To keep temperature-increase below 1.5°C, we therefore need to keep the warming from *everything else* below 0.8°C.

AN AMERICAN IN PARIS

Ex-President Donald Trump caused headlines when he announced that he was pulling the USA out of the Paris Climate agreement in 2017. Due to legalities around the withdrawal, this was only made official on 4 November 2020. One day after the election. The election Trump lost. Oh, the delicious, delicious irony. On his first day in office, President Biden overturned the withdrawal from the Paris climate agreement and the USA is currently re-joining the agreement.

PARIS CLIMATE AGREEMENT

FINANCE

Rich countries will provide $100 billion to developing ones for climate change adaptation by 2020.

ADOPTED THE AGREEMENT

196 countries officially recognising human influence on climate.

DEADLINE

Will come into force by 2020 if signed by 55 countries covering 55% of global emissions.

AMBITION

Every 5 years countries shall revise their emissions reduction targets and measures.

GOAL

Holding the increase in the global average temperature well below

2°C

Pursue efforts to limit the temperature increase to

1.5°C

CLIMATE NEUTRALITY

The balance between emissions and sinks should be reached in the second half of 21st century.

CLIMATE DAMAGE

For the first time ever the Agreement defines climate loss and damage terms but liability and compensation are not mentioned.

CLEAN TECHNOLOGIES

The Agreement urges to speed up clean tech development and international technology transfer.

ROLE OF FORESTS

The Agreement covers saving and increasing forest area in order to capture GHGs from the atmosphere.

CONCLUSION

It's easy to feel overwhelmed by the sheer magnitude of the climate crisis. If you feel sad, or frustrated or downright pissed-off, it's because you understand what is _actually_ happening. We must allow ourselves to surrender to the reality of the crisis to begin the massive task of saving the planet. We are already losing hundreds of species a year, people are losing their homes, their lives and whole communities are being destroyed.

The time to act is now (actually technically it was a good few decades ago so call us when you invent a time machine, yeah?). The climate crisis is not just some inconvenience, it is a major threat and it will get worse the longer we leave it. The COVID-19 pandemic showed us that it is possible for governments to act relatively quickly in response to a crisis – with various degrees of success, consistency and ethics. What we do see, however, is that change, when faced with a huge threat to life, is possible on a huge scale. Climate change is much more existentially threatening than COVID-19 alone. In fact, an increasing occurrence of global pandemics is one of the many consequences of a less robust climate system.

People struggle to take an empathic leap, yet they cannot ignore what is happening in their immediate environment. We have a great deal of difficulty feeling the way we _need_ to feel in order to respond to climate change with immediacy. Our care has to extend beyond ourselves and our loved ones. It has to extend to people in other parts of the world and to the people who will live even in the furthest future. We need to be moved and motivated _every day_ to act in ways that help us to fix the multiple and embroiled problems or racism, injustice, inequality, lack of equity and increased conflict that are part of climate change.

Our failure to deal with climate change is as much to do with what happens within and between us as what is happening to the planet. A society of extreme inequality **WILL. NOT.SAVE.THE.PLANET!** Neoliberalism, austerity and free market capitalism **WILL. NOT.SAVE.THE.PLANET!** Billionaires **WILL.NOT.SAVE.THE.PLANET!** We need a new post-capitalist system that allows us to imagine and implement an economic system that benefits ecosystems and benefits EVERYONE, and the planet we share.

We need everything we've got to save the f**king planet.

RESOURCES

WEBSITES
Climate Signals
https://www.climatesignals.org/
IFL Science
https://www.iflscience.com/
World in Data
https://ourworldindata.org/
Nature https://www.nature.com/

BOOKS
A People's Green New Deal by Max Ajl
Against the Anthropocene: Visual Culture and Environment Today by T. J. Demos
Arts of Living on a Damaged Planet: Ghosts and Monsters of the Anthropocene by Anna Lowenhaupt Tsing, Heather Anne Swanson, Elaine Gan, Nils Bubandt (eds)
Burning Up: A Global History of Fossil Fuel Consumption by Simon Pirani
China on the Edge: The Crisis of Ecology and Development in China by He Bochum
Climate, Capitalism and Communities: An Anthropology of Environmental Overheating by Astrid B Stensrud, Thomas Hylland Eriksen
Decolonizing Nature: Contemporary Art and the Politics of Ecology by T. J. Demos
Don't Even Think About It: Why Our Brains Are Wired to Ignore Climate Change by George Marshall
Losing Earth: The Decade We Could Have Stopped Climate Change by Nathaniel Rich
How to Give Up Plastic: A Guide to Changing the World, One Plastic Bottle at a Time by Will McCallum
Exploring Degrowth: A Critical Guide by Vincent Liegey, Anitra Nelson
Just Transitions: Social Justice in the Shift Towards a Low-Carbon World by Edouard

Morena, Dunja Krause, Dimitris Stevis (eds)
Mining Encounters: Extractive Industries in an Overheated World by Robert Jan Pijpers, Thomas Hylland Eriksen
No One Is Too Small to Make a Difference by Greta Thunberg
On Fire: The Burning Case for a Green New Deal by Naomi Klein
The Ends of the World: Volcanic Apocalypses, Lethal Oceans, and Our Quest to understand Earth's Past Mass Extinctions by Peter Brannen
The Mushroom at the End of the World by Anna Tsing
The Uninhabitable Earth by David Wallace-Wells
This Changes Everything: Capitalism vs. The Climate by Naomi Klein
Silent Spring by Rachel Carson
Staying with the Trouble by Donna J. Haraway
Storming the Wall: Climate Change, Migration, and Homeland Security by Todd Miller
World of Matter by Gavin Bridge, T. J. Demos, Timothy Morton
Where the Water Goes: Life and Death Along the Colorado River by David Owen

PODCASTS & YOUTUBE
A Matter of Degrees Podcast
America Adapts
Climate Conversations: A Climate Change Podcast
Climate Queens
Climate Solutions
Crash Course (YouTube)
Drilled
Environmental Insights Podcast
Floodlines
Generation Green New Deal
Hot Take
How to Save a Planet
Inherited
New Books Network

Outrage + Optimism
Podship Earth
Political Climate
Policy Forum Pod
SciShow (YouTube)
System Reboot
Reversing Climate Change Podcast
The Energy Gang

INTERNATIONAL ORGANISATIONS (GOVERNMENTAL AND NGOs)
This selection of International and National Governmental and Non-Governmental organisations offers just a small cross-section of the work being done. Have a look and explore the list.

INTERNATIONAL
350.org
Agricultural Justice Project
Arab Forum for Environment and Development (AFED)
International Fund for Agricultural Development
The World Health Organization
The United Nations
Biofuelwatch
Bioversity International
BirdLife International
Biomimicry Institute
C40 Cities
CEE Bankwatch Network
Citizens' Climate Lobby
Climate Alliance
Climate Action Network (CAN)
Confederation of European Environmental Engineering Societies
Conservation International
Earth System Governance Project (ESGP)
Earth Charter Initiative
Earth Day Network
Earthwatch
Environmental Defense Fund

RESOURCES

Extinction Rebellion

Fridays for Future & School strike for climate (FFF)

Global Green Growth Institute (GGGI)

Global Footprint Network

Global Landscapes Forum

Global Witness

GoodPlanet Foundation

Great Transition Initiative

Green Cross International

Greenpeace

Indigenous Environmental Network (IEN)

Interamerican Association for Environmental Defense

International Analog Forestry Network

International Marinelife Alliance (IMA)

Intergovernmental Panel on Climate Change (IPCC)

International Rivers

International Tree Foundation

International Union for Conservation of Nature (IUCN)

Julie's Bicycle

La Via Campesina

Rainforest Action Network

Rainforest Alliance

Rainforest Foundation Fund

United Nations Environment Programme (UNEP)

Wetlands International

World Resources Institute

World Wildlife Fund (WWF)

AFRICA

African Conservation Foundation

African Wildlife Foundation

Conservation Alliance of Kenya

Dolphin Action & Protection Group (South Africa)

Earthlife Africa

Eastern Cape Parks (South Africa)

Endangered Wildlife Trust

Green Belt Movement (Kenya)

Groundwork South Africa

Natural Justice: Lawyers for Communities and the Environment (South Africa)

Pan African Climate Justice Alliance (PACJA)

Population Health and the Environment PHE (Ethiopia)

Pragya Kenya

The Okapi Fund for Nature Conservation In the DRC

Ugandan Women's Water Initiative (UWWI)

Wildlife & Environment Society South Africa

Women Environment Project

AUSTRALIA & NEW ZEALAND

Australian Conservation Foundation

Australian Wildlife Conservatory

Bush Heritage Australia

Birds Australia

Clean Ocean Foundation

Conservation Volunteers New Zealand

Greening Australia

Live Ocean

New Zealand Climate Network

New Zealand's Ecological Restoration Network

New Zealand Institute for Environmental Health

Oceans Watch

Wildlife Watch Australia

CANADA

Ancient Forest Alliancet

Canadian Environmental Network

Canadian Youth Climate Coalition

David Suzuki Foundation

Earth Liberation Army (ELA)

Earth Rangers

Nature Canada

Regenesis (non-profit organization)

Sierra Club Canada

Stand.earth

CARRIBEAN & CENTRAL AMERICA

Caribbean Community Climate Change Center (CCCCC)

Red de Información C&T para América Latina y el

Caribe (Network of Science and Technology Information on Latin America and the Caribbean)

La Asociación Costarricense de Turismo Rural Comunitario - ACTUAR (Costa Rican Association of Rural Community Tourism)

Asociación Conservacionista de Monteverde - ACM (Monteverde Conservation Association)

Instituto Nacional de Biodiversidad - INBio (National Institute for Biodiversity)

Asociación de Geógrafos Profesionales de Panamá (Association of Professional Geographers of Panama)

CHINA

Clean Air Network (Hong Kong)

Clear the Air (Hong Kong)

China Youth Climate Action Network (CYCAN)

Friends of the Earth (Hong Kong)

Friends of Nature

Green Camel Bell

Green Council (Hong Kong)

Green Power (Hong Kong)

Lights Out Hong Kong

Society for Protection of the Harbour (Hong Kong)

The Conservancy Association (Hong Kong)

The Climate Group (Hong Kong)

EAST AND SOUTHEAST ASIA

Agkor Centre for Conservation of Biodiversity

Alyansa Tigil Mina (Philippines)

Borneo Orangutan Survival Foundation (Indonesia)

CANSEA Climate – Climate Action Network Southeast Asia

Climate Action for South East Asia

Climate Change Working Groups Vietnam

Free the Bears Fund (Cambodia)

Friends of the Earth

Gili Eco Trust (Indonesia)

Japan Climate Initiative

The Indonesian Forum for Environment

MCD Marine Conservation and Community Development (Vietnam)

Nature Society (Singapore)

Korean Federation for Environmental Movement

Organisation for the Preservation of Birds and their Habitat (Indonesia)

Save Cambodia's Wildlife

Sibuyanons Against Mining (Philippines)

Partnerships in Environmental Management for the Seas of East Asia (PEMSEA)

Yayasan Merah Putih (Indonesia)

EUROPE

Asociacion pola defensa ria (Spain)

Bellona Foundation (Norway)

Client Earth

Climate Action Network Europe (CAN-Europe)

Coastwatch Europe

European Association of Environmental and Resource Economists

Gluaiseacht (Ireland)

Green Artvin Association -Yeşil Artvin Derneği - (Turkey)

Green Warriors of Norway

Health and Environment Alliance (HEAL)

Irish Peatland Conservation Council (IPCC)

Legambiente (Italy)

Macedonian Ecological Society

Mediterranean Association to Save Sea Turtles (MEDASSET)

Nature and Youth (Norway)

Notre Affaire a Tour (France)

Norwegian Society for the Conservation of Nature

The Climate Coalition (UK)

Quercus (Portugal)

Rainforest Foundation UK

Rientrodolce (Italy)

Zero Emission Resource Organisation (Norway)

INDIA

Conserve

Delhi Greens (NGO)

Environmentalist Foundation of India

Pasumai Thaayagam TNPT

Poovulagin Nanbargal

Pragya India

Vindhyan Ecology and Natural History Foundation

Wildlife Trust of India (India)

PACIFIC REGION AND POLYNESIA

Asia Pacific Adaptation Network (APAN), Asia and Pacific Region

Climate Change and Communications (Tonga)

Micronesian Shark Foundation, Palau

Ministry of Environment, Conservation and Meteorology (Solomon Islands)

National Trust of Fiji, Fiji

Palau Conservation Society

Palau Protected Areas Network Fund

Te Ipukarea Society (Cook Islands Environmental NGO)

Te Mana o Te Moana (French Polynesia)

The Tenkile Conservation Alliance (Papua New Guinea)

RUSSIA

Biodiversity Conservation Centre (BCC)

Bellona (Russia)

Friends of the Earth Russia

WWF Russia

SOUTH AMERICA

Asociación Argentina de Energías Renovables y Ambiente - ASADES (Argentine Association of renewable Energies and Environment)

Fundación PROINPA

Associação dos Geógrafos Brasileiros (Association of Brazilian Geographers)

Instituto de Ecología y Biodiversidad - IEB (Institute of Ecology and Biodiversity in Chile)

La Fundación Grupo HTM (Hábitat, Territorio y Medio Ambiente) (Foundation Group HTM (Habitat, Territory, and Environment)

USA

Climate Collaborative

Climate Justice Alliance (CJA)

Earthjustice

Environmental Working Group

Fund for Wild Nature

Inter-Tribal Environmental Council

National Oceanic and Atmospheric Administration

Justice for Migrant Women

Rainforest Foundation US

Sunrise Movement

Yellowstone to Yukon Conservation Initiative

WESTERN AND CENTRAL ASIA

Association for the Conservation of Biodiversity of Kazakhstan (ACBK)

Bangladesh Environmental Lawyers Association

Environmental Fund for Lebanon (EFL)

Himalayan Wildlife Foundation (Pakistan)

Israel Union for Environmental Defense

Pakistan Environmental Protection Agency (Pak-EPA)

Palestinian Environmental NGOs Network

Persian Wildlife Heritage Foundation (Iran)

Project Green Oman

Saudi Environmental Society (Saudi Arabia)

Society for the Protection of Nature in Israel

UNDP in Afghanistan

Uzbekistan Society for the Protection of Birds

Zalul Environmental Association (Israel)

CREDITS

Aquir/Shutterstock: 13, 19, 37, 63, 71, 81, 99, 109, 136, 161, 175

P. 21 Gettelman A., Rood R.B. *Essence of a Climate Model.* Demystifying Climate Models. Earth Systems Data and Models, vol 2, 2016. Springer, Berlin; Heidelberg. Accessed at: https://doi. org/10.1007/978-3-662-48959-8_4

P. 24 Wanous, S. *Use psychology for better climate communications.* Citizens Climate Lobby. December 11, 2018, Accessed at: https:// citizensclimatelobby.org/use-psychology-for-better-climate-communications/

P. 35 Tapon, F. *Earth's History Compressed in One Year,* Wanderlearn with Francis Tapon, n.d. Accessed at: https://francistapon.com/Travels/ Continental-Divide-Trail/Earth-s-History-Compressed-in-One-Year

P. 41 Pokar, M. *Oldest ever ice core promises climate revelations,* The New Scientist, September 8, 2003. Accessed at: https://www.newscientist.com/ article/dn4121-oldest-ever-ice-core-promises-climate-revelations/

P. 44-5 Scott, M and Lindsey, R. *What's the hottest Earth's ever been?* Climate. gov, June 18, 2020. Accessed at: https:// www.climate.gov/news-features/ climate-qa/whats-hottest-earths-ever-been

P. 51 Mann, M. E., Bradley, R. S., and Hughes, M. K. *Northern Hemisphere Temperatures During the Past Millennium: Inferences, Uncertainties, and Limitations,* AGU GRL galley style, v3.1, 14 February 1994. Accessed at: http://www.meteo.psu.edu/holocene/ public_html/shared/research/ONLINE-PREPRINTS/Millennium/mbh99.pdf

P. 68 Shendruk, A. *Charting the world's sixth mass extinction,* Maclean's, June 29, 2015. Accessed at: https:// www.macleans.ca/society/science/ infographic-charting-the-worlds-sixth-mass-exinction/

P. 73 © The City University of New York. As seen in: van Gerven, M. *Why True Agile Transformation Requires Apex Predators,* Organize Agile, July 30, 2019. Accessed at: https://www. organizeagile.com/update/why-true-agile-transformation-requires-apex-predators/

P. 91 © Climate Watch, The World Resources Institute. As seen in: Ritchie, H. *Sector by sector: where do global greenhouse gas emissions come from?* Our World in Data, September 18, 2020. Accessed at: https://ourworldindata. org/ghg-emissions-by-sector

P. 100 Enerdata. *Leaders in Renewable Energy,* Visual Capitalist, October 31, 2018. Accessed at: https://www. visualcapitalist.com/global-transition-to-green-energy/

P. 107 © Oil Spill Intelligence Report. As seen in: Black, R. *Gulf oil leak: Biggest ever, but how bad?* BBC, August 3, 2010, Accessed at: https://www.bbc.co.uk/ news/science-environment-10851837

P. 113 © InfoDiagram.com. As seen in: *Principles of Urban Sustainability.* Finding Sustainable Solutions to our Transportation Problems, February 11, 2018. Accessed at: https:// us130urbansustainability.wordpress. com/2018/02/11/finding-sustainable-solutions-to-our-transportation-problems/

P. 121 © Cowspiracy. As seen in: *Movie Reviews Cowspiracy: A Film Review,* Wild Lens, July 3, 2017. Accessed at: http://www. wildlensinc.org/cowspiracy-film-review/

P. 133 © Aaron McConomy. As seen in: Denchak, M. *Hurricanes and Climate Change: Everything You Need to Know,* NRDC.org, December 3, 2018. Accessed at: https://www.nrdc.org/stories/ hurricanes-and-climate-change-everything-you-need-know

P. 139 © Statista, Associated Press. As seen in: Zarrell, M., *Using US map to examine scale of massive Australia wildfires,* ABC News, January 7, 2020. Accessed at: https://abcnews.go.co/ International/us-map-examine-scale-massive-australia-wildfires/ story?id=68102703

P. 141 © Ellen Macarthur Foundation and A New Textiles Economy & EEA Europa. *Luft und schiffsverkehr im Fokus,* 2016. As seen in: *The Environmental Impact of the Fast Fashion Industry,* SANVT, March 12, 2020, Accessed at: https://sanvt.com/ journal/environmental-impact-of-fast-fashion-infographic/

P. 142 *Why Fashion?* Ocean Tee, n.d. Accessed at: https://oceanteegolf.com/ why-fashion/

P. 153 *The Lifecycle of Plastics,* The World Wildlife Fund Australia, June 19, 2018. Accessed at: https://www.wwf. org.au/news/blogs/the-lifecycle-of-plastics#gs.ylx3c3

P. 159 *The Great Pacific Garbage Patch,* The Ocean Cleanup. Accessed at: https://theoceancleanup.com/great-pacific-garbage-patch/

P. 161 © World Wildlife Fund as seen in Olenick, L. *The Cautionary Tale of DDT – Biomagnification, Bioaccumulation, and Research Motivation,* Sustainable Nano: A blog by the NSF Center for Sustainable Nanotechnology, December 17, 2013. Accessed at: https:// sustainable-nano.com/2013/12/17/ the-cautionary-tale-of-ddt-biomagnification-bioaccumulation-and-research-motivation/

P. 165 Strauss, K. and Lindebo, E. *Investing in the transition to thriving EU waters: A visionary new framework,* Environmental Defense Fund, July 28, 2014. Accessed at: http://blogs.edf.org/ edfish/page/29/?redirect=edfish

P. 173 *Air Pollution Infographics,* World Health Organisation, n.d. Accessed at: https://www.who.int/airpollution/ infographics/en/

P. 174 *Air Pollution Infographics,* World Health Organisation, n.d. Accessed at: https://www.who.int/airpollution/ infographics/en/

P. 176-5 Das, S. and Dutt, F. *Eco-Urban Planning & Design for a futuristic vision of Shanghai,* October 2012. Accessed at: https://www.researchgate.net/ publication/309592883_Eco-Urban_ Planning_Design_for_a_futuristic_ vision_of_Shanghai

P. 180, *Blood Lead Levels in Children Aged 1-5 Years- United States, 1999-2010,* Centres for Disease Control and Prevention, April 5, 2013, 62 (13); p. 245-248. Accessed at: https://www. cdc.gov/mmwr/preview/mmwrhtml/ mm6213a3.htm

P. 187 *Infographics: COP21 results in historic agreement.* Climate Russia, December 16, 2015. Accessed at: http:// climaterussia.org/policy-and-finance/ infographics-cop21-results-in-historic-agreement